PRAISE FOR STEVE GROFF AND *THE FUTURE-PROOF FARM*

"Every part of the story that Steve tells in this book rings true to me. I've seen so much of it with my own eyes since I first met Steve planting no-till potatoes in his kitchen garden with his now grown son on his knee. I'm a professor, but Steve has been one of my greatest teachers. For twenty-five years I have been taking my college students up to Cedar Meadow Farm to learn from Steve. So, it is very gratifying to see his wisdom and insight now available in a book for all to read."

—**RAY WEIL**, *professor of soil science, University of Maryland*

"If you want to be successful and make a lot of money farming, you should only accept advice from farmers who have achieved the same goals you are striving for. The most successful farmers think differently than most and see the world differently. This book is your chance to glimpse how a very successful farmer and farming coach thinks differently about agriculture and food production."

—**JOHN KEMPF**, *founder of Advancing Eco Agriculture*

"Steve is truly a 'pioneer' who has weathered a lot of storms in his life and has come out strong. He has been a mentor for me, has helped build my confidence to be a leader in the promotion of soil health, and has helped change the mindset of many in the field of cover crops and soil health."

—**JIM HERSHEY**, *owner, Hershey Farms LLC, and president of the Pennsylvania No-Till Alliance*

"Farmers listen to other farmers. Innovative farmers with credibility are the most effective mentors. Steve has Cedar Meadows Farm as evidence that he 'walks the talk,' sharing his experiences in making the soils richer in carbon, healthier, and more productive and profitable."

—DON REICOSKY, *USDA-ARS retired soil scientist, North Central Soil Conservation Research Laboratory, Morris, Minnesota*

"Steve Groff may be the Cover Crops Coach and the father of the Tillage Radish, but now we get to add Explainer-in-Chief; this book is a clear, concise, and useful guide to practical farming for the future of good food."

—DAN BARBER, *author, chef, and co-owner of Blue Hill in Manhattan and Blue Hill at Stone Barns in Pocantico Hills, New York*

"From my own medical viewpoint, it is becoming increasingly clear that our gut biome is intimately related to our overall health, in which the quality of the food we eat is key. Steve addresses the high nutritional value of food produced by regenerative techniques. This book is well written, informative, interesting, and down to earth. It is gently sprinkled and seasoned with faith quotes, Steve showing his credit and appreciation to his creator God."

—DR. LOREN HELMUTH, MD

"I got to know of Steve's work on cover crops through YouTube videos. In November 2019 I had the great opportunity of sharing his knowledge during a whole week with his visit to Argentina and Uruguay. Now after reading his WAKE-UP CALL book, I have learned more about his farming awareness of the environment as well as the social issues of farming healthier."

—MARIA TERESA, *farmer and leader of Suelos y Sistemas (Soils and Systems) in Argentina and Uruguay*

"*The Future-Proof Farm* is a wake-up call to us as agricultural producers. It is time to open our eyes and meet our new bosses and start delivering what they are demanding. I appreciate how Steve clearly lays out that the future relevancy of our farms will depend on the regeneration of both our mindsets and our soils."

—KEITH BERNS, *co-owner and sales manager, Green Cover Seed*

"I enjoyed the book very much. It's engaging, full of that original Groff spirit with which he approaches everything he does, a nice blend of personal and agriculture, a nice blend of ecology and economics, and it's practical/useful stuff. I met Steve Groff twenty years ago when we teamed up to provide locally grown vegetables to supermarkets in Philadelphia. Steve is one of those people who can sense the winds of change long before the rest of us do. He was using cover crops, experimenting with no-till, and applying tons of organic matter to his fields when the language of sustainable agriculture was only beginning to enter common usage. Now we get to benefit from his decades of learning, experimentation, mistakes, and proven accomplishments. Whether you're a farmer or someone else who wants to know more about regenerative farming, this book shows how cover cropping and related practices are beneficial for the soil and profitable for the farm business at the same time."

—MICHAEL ROZYNE, *founder of Red Tomato and cofounder of Equal Exchange*

"I think of Steve Groff as an early cover crop pioneer who helped me with my own changing mindset. He continues to lead the way toward more resilient soils with *The Future-Proof Farm*."

—JAY FUHRER, *soil-health specialist, retired, Natural Resources Conservation Service, United States Department of Agriculture*

"Like cover crops, no-till, and the Tillage Radish that he has promoted, Steve Groff is a deep-rooted farmer. His operation, through thirty years of practice, has been his teaching room, his lab, but also the filter between theory and real life in the field. His experience, curiosity and openness helped him to become an advocate of cover crops and no-till. In this respect, he spent years educating farmers and others on why we must pay attention to what nature is telling us. His time sharing—in the US, but also across the world—not only helped him to master his practical knowledge but also brought him a wider and wiser view and analysis of farming, agriculture, and the world in general. Through a comprehensive approach between agriculture history, his own farm practices, his personal development, and large general statements, he explains that everything comes down to a simple principle: better soil leads to better food, which promotes better health! Whether we like it or not, the consumer is our boss, and big food and fiber companies have already made some great moves toward farming practices and farmers who take better care of their soils. This book is finally a very positive message to keep moving in order to start harvesting the fruit from the seed we have been planting for so many years."

—FRÉDÉRIC THOMAS, *French farmer, founder of* TCS *magazine (Soil Conservation Tillage mag) and fifteen-year chairman of the farmers' association BASE (Biodiversity, Agriculture, Soil and Environment)*

"After two decades as a consulting wildlife biologist, I've never been so optimistic about restoring soil through regenerative wildlife agriculture. It's hard to imagine the library of a wildlife 'food plotter' without a copy of *The Future-Proof Farm*. Steve Groff, cover crop pioneer and innovator of the roller crimper, does a fantastic job discussing the practices of building a more resilient recreational property. The agricultural principles of biomimicry promoted in this book are having long-lasting, positive impacts on wildlife populations and their habitats! Groff delivers a road map, with remarkable clarity, for each of us to follow during our journey!"

—JASON R. SNAVELY, *CWB, CEO, Drop-Tine Wildlife Consulting, www.droptinewildlife.com*

"Several years ago we produced an award-winning PBS documentary about Steve Groff and his cutting-edge no-till farming techniques. Steve's new book, *The Future-Proof Farm*, goes well beyond what we covered and is a must-read for anyone interested in saving the planet."

—**HAL WEINER**, *executive producer,* The Journey to Planet Earth *television series*

"Steve's book lays out in full detail the impact no-till, cover crops, and regenerative ag will have in meeting the demand for higher-quality, better-tasting, and environmentally correctly produced food. It is a wake-up call as to how consumer habits and changing food tastes will impact the way you farm. Steve maintains that continuing to produce crops and livestock the way you're doing today will make your farming operation obsolete in a few years. Gear up for change, as the way you farm today won't cut it five years from now."

—**FRANK LESSITER**, *editor,* No-Till Farmer

"It is only fitting that the Cover Crop Coach should package his knowledge into helping us all 'future-proof' our farms. Steve's years of innovation have laid the groundwork for the next generation of innovators. I am excited to be one of the young guys to pick up the torch, rally the troops, and carry on the legacy of Steve and other innovators who have laid out the path to a sustainable future in agriculture."

—**MITCHELL HORA**, *seventh generation Iowa farmer and founder/ CEO, Continuum Ag*

"Steve Groff has been a tireless proponent and promoter of the use of cover crops as an integral part of no-till farming systems for longer than anyone. It is similar to the country and western song: he was 'country when country wasn't cool.'"

—**DWAYNE BECK, PHD**, *manager, Dakota Lakes Research Farm, South Dakota State University*

"Steve Groff has traveled around the world to speak as a leading farmer expert on the topics of cover crops, no-till, and regenerative agriculture. In this book, he describes his personal journey with farming, including why he adopted cover crops and no-till, and why a regenerative approach to agriculture will better support soil health while meeting consumer expectations in the coming years. I first visited Steve Groff's farm in the mid-1990s, at a time when he was one of the only farmers to be experimenting with cover crops combined with no-till. I've been back to his farm a few times over the years, and as he describes so well in this book, his progress toward regenerative agriculture has been remarkable. Any farmer or individual interested in food production can benefit from reading about Steve's experiences not only with his own farm, but in traveling around the world to advise farmers and companies on cover crops, no-till, and regenerative approaches."

—ROB MYERS, PHD, *national liaison on cover crops and soil health, USDA-SARE*

"I have seen Steve speak several times, and I learned something every time. He is one of my mentors, and I wish he had written this book fifteen years ago so I would have been able to get everything I need to know about cover crops in one spot. The book would have saved me from making a lot of the mistakes that I have made. It was great to read his story while learning through all his experiences"

—TREY HILL, *farmer, Harbor View Farms*

"As a stream and river ecologist, I love and respect Steve's simple message: farming with no-till and cover crops conserves water and soil health, leading to less pollution, more nutritious food, and increased farm profit, which is what consumers want. His book is timely, further advances sustainable agriculture, and convinced me that cover cropping can and does lead to clean fresh water, which is what I want!"

—BERN SWEENEY, *distinguished senior research scientist, Stroud Water Research Center*

"It is not an exaggeration to say that Steve inspired me to start my cover crop business. He has been a great mentor along the way, and his book is just like an honest conversation with him: you get practical, firsthand, and ready-to-use advice from 'the Coach.' I'd recommend *The Future-Proof Farm* to every cover cropping farmer and consultant who wants to get it right for the first time."

—ZSOMBOR DIRICZI, *founder, Démétér Biosystems, Hungary*

"Our survival on this planet is dependent on the success of our farmers and the health of our soil. Steve's book is the perfect resource for farmers to get a handle on their changing role. The work he's doing with hemp, especially his leadership role in the community, is laying the groundwork for the farming of the future. This is an inspiring book!"

—ERIC HURLOCK, *digital editor,* Lancaster Farming Industrial Hemp Podcast

"This is an excellent, well-written book that covers the need to change, and how to transform the world's cropping practices, through the eyes of a farmer who has done it."

—COLIN SEIS, *farmer, rancher, agricultural consultant, and educator, Australia*

"I fondly remember my visit to Steve Groff's farm in the late 1990s. He was wearing a cap with the words 'Soil Is Meant to Be Covered.' I was very impressed by this simple statement, which brings it to the point: the soil must never be bare if you are looking for agricultural sustainability. A good farmer should never be able to see his soil, always protecting it, like an umbrella, from pounding raindrops, intense sunlight, and other weather extremes. As long as I've known Steve, his first priority, wholeheartedly, is to be a good steward of the soil. This book is full of actual, timely, and useful examples that show that cover crops are not costly, but that they will pay."

—ROLF DERPSCH, *international conservation agriculture consultant, Germany*

THE
FUTURE-PROOF
FARM

STEVE GROFF

THE FUTURE-PROOF FARM

CHANGING **MINDSETS** IN A CHANGING **WORLD**

Advantage

Published by Advantage, Charleston, South Carolina.
Member of Advantage Media Group.

ADVANTAGE is a registered trademark, and the Advantage colophon is a trademark of Advantage Media Group, Inc.

Printed in the United States of America.

10 9 8 7 6 5 4 3 2

ISBN: 978-1-64225-186-9
LCCN: 2020911752

Book design by Carly Blake.

This publication is designed to provide accurate and authoritative information in regard to the subject matter covered. It is sold with the understanding that the publisher is not engaged in rendering legal, accounting, or other professional services. If legal advice or other expert assistance is required, the services of a competent professional person should be sought.

Advantage Media Group is proud to be a part of the Tree Neutral® program. Tree Neutral offsets the number of trees consumed in the production and printing of this book by taking proactive steps such as planting trees in direct proportion to the number of trees used to print books. To learn more about Tree Neutral, please visit **www.treeneutral.com**.

Advantage Media Group is a publisher of business, self-improvement, and professional development books and online learning. We help entrepreneurs, business leaders, and professionals share their Stories, Passion, and Knowledge to help others Learn & Grow. Do you have a manuscript or book idea that you would like us to consider for publishing? Please visit **advantagefamily.com** or call **1.866.775.1696**.

I dedicate this book to all those who influenced me on this agriculture journey of innovation by teaching me how to be a better steward of God's creation.

But most importantly to my family, who have supported my efforts over the decades. Cheri, you've always said you "married the farmer and not the farm," and that's been okay with me. Dana, I know you treasure the memories of farm life. Lauren, well, thanks for introducing us to musical theater! David, the heir apparent to the farm, I couldn't do all this without your commitment to taking care of this piece of earth that we've been privileged to manage.

CONTENTS

FOREWORD

When Steve Groff introduced himself to me at a Missouri soil health conference where we were both speaking, I knew immediately that I wanted to learn more about his approach to farming responsibly in this world of rapid change. Friendly and jovial, Steve stands out in a crowd by proudly wearing a signature felted hat. He is generous with his passion and knowledge as he offers sage advice on the intersection between agriculture and the preservation of soils. And, as I have had the opportunity to get to know Steve more over the years, it is obvious that his respect for nature and the land drives his commitment as a steward.

I have long admired people with such a commitment to stewardship—John Muir, Aldo Leopold, Helen and Scott Nearing. Their words and philosophies inspire a connection with nature that lives outside ourselves and humbles the human experience. From an early age I aspired to better understand the steward's commitment and devote myself to learning from the marvels of nature. For four summers, in between college courses, I was a canoe guide in the wilds of northern Maine, leading river trips for a week or longer into the backcountry for fellow adventurers. Combating the black flies, we would return bearded, wet, tired, and hungry and with sublime experiences and memories—like sidling up to a group of loons in full chorus under the moonlight in the middle of Grand Lake Matagamon. Later I worked with indigenous tribes in remote villages of Alaska, teaching swimming skills in a land where the water claims so many lives.

Those early days are forever etched in my heart. It was a time of deep connection to nature, but also with a yearning to share with the world what I had experienced. My next level of stewardship commitment was to bring these ineffable nature lessons into the human world. In reality, these two worlds are connected. Nature does not discriminate between the animal kingdom and human society. We cannot isolate ourselves from something that is part of us. We can make a conscious choice, as a society and species, to evolve from a force of destruction to a symbiotic collaborator with the natural world.

At Duke University, I added some credentials to my passion, graduating with a master's in environmental management. In radical collaboration my classmates and I sought to tackle wicked problems by exploring all possible solutions, whether from policy, economics, or governance. I developed an interest in and excitement for the private sector because I saw the speed at which ideas could come to fruition.

In the years since, I have found my place as director of sustainability for Wrangler, a brand that most everyone recognizes and that symbolizes for many the values of courage and connection to the land. This is a company where I can fulfill my stewardship commitment. With a dedication to protecting the resources of our planet, Wrangler offers "tough denim, gentle footprint, responsible action." We strive to team with farmers worldwide who produce cotton in a way that protects and regenerates the soil. The earth provides for us in so many ways. We at Wrangler recognize our duty to give back to it.

I have a deep respect for our agricultural community. Farmers bear an awesome responsibility to feed and clothe us. They do it faithfully, year in and year out, despite a host of hardships and risks. The weather and pests can ravage their harvest. The economy and shifting markets can evaporate their profits. It's not an easy way to make a living, but most would prefer to do nothing else. Farming is more than a business. It's a way of life, one worth preserving—and farmers are finding that a good way to accomplish that is to take steps to preserve and restore their primary resource, the soil. Understanding soil is like understanding nature, with all of its nuances and nebulas of deep ecological interdependencies, microscopic alien creatures with Latin names; and the next frontier of regenerative agriculture is that opportunity for a symbiotic collaboration with nature.

It is my privilege to work closely with cotton producers who are dedicated to the stewardship of the land and have made strides toward farming responsibly. Their ability to adapt is critical. And that is why I am privileged to know Steve Groff, a true man of the land and a born educator. He is teaching farmers around the world how they can adapt and thrive, taking meaningful steps today to accelerate their own stewardship commitments. Steve is always welcome at my campfire. As a champion for cover crops, no-till farming, and regenerative agriculture, Steve has dedicated years to cultivating the new mindset that will future-proof our farms.

ROIAN ATWOOD
Senior director of sustainability, Wrangler

WHAT IS A COVER CROP?

N ow and then when I tell new acquaintances that I am an authority on cover crops, I can see by the look on their faces that our conversation needs to start out with some basics. They have a question or two for me at the outset. Such as: "Steve, what is a cover crop?"

Originally, farmers planted cover crops to "cover" the land during the off-season of their cash crops. The aim was to prevent erosion by water and wind. Today, cover crops are also valued as a natural means of fertilizing the soil, building organic matter, and controlling weeds and pests, all for the benefit of the soil's health. Generally, cover crops are planted for those reasons and not for the primary purpose of selling them for a profit.

Typically, farmers plant a cover crop in the fall, after they have harvested the cash crop, to protect the soil over the winter. With that protection in place, the soil stays in the field instead of eroding into the waterways along with its nutrients and any pesticides and fertilizers that it might contain. The soil remains on the farm where it belongs. Sometimes, farmers also plant cover crops during the summer, such as after the harvest of winter wheat in early July. In that case, the crop protects the soil from baking in the sun.

Cover crops help the soil to function as nature intended, sort of the way that our skin covers and regulates our bodies. Skin protects us from the elements, regulating our body heat and moisture, warming us and cooling us. We further cover ourselves with clothing appropriate to the season. The soil needs a "skin," too, to protect it from the wind and rain and to regulate its heat and moisture so that it can function properly. God didn't intend for the soil to be laid bare and barren. It needs a cover. That is what nature provides when weeds grow on unattended land. When we farm that land, that is what cover crops can provide.

> # Nothing in the definition of *cover crop* says "must not be grown for profit."

In essence, a cover crop keeps farmland healthy by holding the soil in place between the plantings of cash crops. It cools the soil in the summer, warms it in the winter, and enriches it with organic matter and natural nutrients.

Theoretically, any plant could be used as a cover crop, because the soil doesn't know whether the farmer intends to harvest it for profit. If he plants it for the purpose of providing cover, then it is a cover crop.

That doesn't mean the farmer can't make money off it. Nothing in the definition of *cover crop* says "must not be grown for profit." For example, with enough rainfall, a summer cover of sorghum and Sudan grass might shoot up to three or four feet tall. With that kind of growth, I might decide to cut and bale it in September and sell it to my Amish neighbors to feed their cows—and I still would get a foot or more of regrowth after that to protect the soil through the winter months.

Nor does anything in a definition of *cash crop* say "must not be grown for cover." Some plants that typically would be harvested for profit can also be used in the context of a cover crop. I developed daikon-type radishes that certainly were edible as a vegetable but that we planted for cover, without the intention of harvesting them to eat. I also plant an oat variety that does well for cover cropping, but I grow it to produce seeds that I sell to other farmers. Those all are examples of cover crops that can double as a cash crop, or vice versa.

Today, I would say that eighty to a hundred types of plants have been used as cover crops, including those that can be grown as cash crops as well. And there are other ways that they can serve a dual purpose. Sometimes farmers will graze their cattle on a cover crop or cut the growth in the spring for silage. In that case, not only do the crops protect the soil, but they also sustain the livestock, which in turn add nutrient-rich manure to the soil and produce meat for

There's nothing more brilliant than a blooming cover crop of Crimson Clover!

the dinner plate. That's one of many variations in how cover crops can be used to enhance soil health.

Three main classes—grasses, legumes, and brassicas—make up most of the cover crops in use today. The most popular of the grasses is cereal rye because of its versatility: it can be planted early or late, and it is able to survive harsh conditions. Triticale and annual rye grass are other common ones. The most popular of the legumes, which are nitrogen-producing plants, are hairy vetch and crimson clover. Among the brassicas, radishes and oilseed rape (canola) are a few of the more common ones.

Cover crops can be used in agriculture on any scale. In the United States, some of the major cash crops by acreage are corn, soybeans, wheat, and cotton—and those commodities present a huge opportunity to make use of the environmental and economic benefits of cover crops. The numerous vegetable producers and other growers also can benefit greatly. Wherever and whenever farmers desire to protect and improve their soil, cut down on chemical costs, increase their yields, and build for a better future, cover crops will serve them well.

Crimson clover, hairy vetch (purple flowers), and cereal rye are some of the most popular cover crops farmers use.

INTRODUCTION

ROOTS OF A PASSION

ROOTS OF A PASSION

It was 1982, the year that I graduated from high school, when I took an interest in no-till farming. I wish I could say that I wanted to do my part to protect the soil and become a champion for the environment, but I confess that I wasn't exactly out to save the Chesapeake Bay back then. Mostly I just wanted to get rid of those confounded erosion ditches that took so much time to fill each year at harvest time. Why do all that work when there was a better way? I felt the same way about going to college.

In the years since, I've gotten quite an education here at Cedar Meadow Farm. I was born for this. Here I grew up, and here I stay. What I have learned comes from decades working under the sun, not from four years buried in textbooks. I'm not against book learning—our house has plenty of good reading on the shelves—but I strongly believe that we grow mostly by experiencing, connecting, and sharing. A fellow can learn a lot by listening and observing. Whenever I want to figure out how to do something better, I get to know folks who already are doing it, watch how they do it, and ask who taught them to do it. That's a good way to learn to farm better. I don't know a better way to learn to do anything.

As you soon will see in the chapters ahead, I have a message to share that my fellow farmers must hear if we want our way of life to endure into the next generation and beyond. Our world is changing swiftly in so many ways, and farmers must either adapt or die out. That is not meant to sound hopeless. If we listen to what the people who buy our products are telling us, we can future-proof our farms. We have an opportunity to thrive as never before while becoming better caretakers of this good earth that God has given us.

This is my "happy place"—in the field, kneeling on the ground, surrounded by eager learners. Discussing my Tillage Radish in the country of Hungary.

I have continued to meet a lot of smart and dedicated people as I have expanded my horizons coast to coast, across the seas, and wherever people are willing to hear about the environmental and economic value of putting away the plow and planting cover crops. I am humbled that so many esteemed professionals and business-people are interested in hearing the essentially simple message from this third-generation farmer from Pennsylvania. They want to get to know someone who is doing it, who is making it work on his own acres—and what good is information unless it is shared?

I deeply respect people who have attained master's and doctorate degrees and such. I also deeply respect those who walk their land and stoop to sniff the hay and feel the soil, then rise to scan the sky. In a college classroom you can learn a lot about farming. It's out in the fields and up on the tractor where you learn to farm. It has been my pleasure to meet a good many people who are comfortable in both settings and who are determined to do things the right way.

Which takes me back to 1982 and those annoying ditches. I had been farming long enough to get sick and tired of filling in those ruts so that we could get our equipment into the fields to harvest the crops. They were more like trenches, sometimes a couple of feet deep. I thought that was just wrong—not because our soil was washing away, which it was, but because I was wasting hours shoveling dirt just so I could get some real work done. It's not that I was arrogant or uncaring, but at the time I was thinking about efficiency, not sustainability. I didn't yet understand how the farmer's traditional plowing methods had been hurting the environment and picking his pocket.

That perspective, or lack of one, would soon change. After trying out no-till on a few fields, I began getting a feeling at the soul level that this was the way we should be farming. I started to see the benefits of keeping the soil intact. The soil, which to farmers like me is everything, was staying in place where God had put it. It wasn't getting swept downstream with every storm. I was a teenager with a lot to learn, but I knew that we were on to a good thing.

At the same time, we had been planting cereal rye as a cover crop to some extent, but not every year and only when it was convenient. The rye was another way of controlling erosion between the plantings of our cash crops. Once I began to see the impressive benefits of no-till, though, I started to wonder whether we would need cover crops at all after giving the soil a decade or so to recover from the plow.

I was still wondering that in 1995, when I met Dr. Ray Weil of the Department of Environmental Science and Technology at the University of Maryland, a man who also works directly with villagers and indigenous people in Africa and Mexico on a variety of soil, water, and crop issues. I asked him whether no-till alone might be sufficient to ensure soil health. "I've been asking the same thing," he told me. "Would you like to join me in some research?"

That year, we began the cover crop experiment, in which we did controlled plantings in successive years for comparison to other fields without cover crops. Otherwise, the fields were farmed the same way. Four years later, after the harvest of 1999, I had my answer. That was a very dry year in Lancaster County, and when I compared the yields from the fields, I knew that I would never turn back from cover crops.

> That was a very dry year in Lancaster County, and when I compared the yields from the fields, I knew that I would never turn back from cover crops.

The land where I had planted cover crops those four years were now producing twenty-eight more bushels of corn per acre. The organic matter was richer, the nutrients were denser, and the overall soil health was better in those fields. *Yes, cover crops do pay*, I concluded, and I resolved to incorporate them everywhere that I could on my farm— and that was my mission as we headed into the new millennium.

About the same time that I was beginning the cover crop experiments, I began working with the Pennsylvania Association for Sustainable Agriculture, where I became a board member. I was conducting a basic experiment using lower-rate herbicides, and I got the idea to invite folks to a little field day here at Cedar Meadow Farm. It was the first time that I had done that, and nearly thirty people showed up. I thought it was awesome that people would want to come to see my farm. After that, I began holding field days every year. Soon my research was being showcased in farm publications. I was developing a reputation as a pioneer and an innovator.

We've hosted dozens of field days like this at our farm over the past couple of decades.

As a result of that first field day at Cedar Meadow, the association asked me to talk about it at its annual winter conference. Never before had I spoken in public—but when I got up in front of those folks, I found that I was comfortable in that role. My passion overcame my fears. In any case, I figured, this would be the last time I did such a thing. Then came a series of requests. Those who heard me approached to ask whether I might be willing to give other talks at other meetings, and when I spoke at those gatherings, I soon got even more requests.

It all snowballed. With my reputation for innovation, I found myself in even more demand to appear at conferences. I can't say that I had ever imagined becoming a public speaker, but I was up for the challenge. Over the years I have worked to hone my craft, reading books and watching videos on how to present oneself convincingly so the audience will remember the message. (One book, *Everyone Communicates, Few Connect,* by John Maxwell, particularly resonated with me.) I found a lot of good tips on TED Talks. In short, I discovered a new me. I enjoyed reaching out and sharing what I knew and comparing notes with colleagues to help us all be the best we could be.

I have remained active in the speaking circuit so that I can spread the word personally. To back that up, I also developed what I consider to be the world's largest library of farmer-produced webinars, over a hundred and fifty of them. I have three Facebook accounts directly related to cover crops, soil health, and, most recently, CBD hemp. An

*I was interviewed by Hungarian media at the first ever
soil health conference in Hungary.*

online search of my name will find numerous articles and YouTube videos, and I am active on Twitter with over 10,000 followers.

Nonetheless, the internet is not my primary platform. All that online and social media activity is a by-product of what I do. The real work comes handshake by handshake, whether I'm speaking to ten people gathered at an Amish farm or a thousand or more at a national conference.

I do a dozen or more meetings each year for the local Amish community, teaching them how to use cover crops and no-till.

As I write this, a quarter century after my first speaking engagement, I am still going strong. I feel that I have found a purpose that combines my experience working the land with my desire to forge connections and educate. My greatest satisfaction comes from inspiring others to go out and make a difference. I find it gratifying to see others succeed and to know that I played a role in their progress. My mission is to put people together for the purpose of taking good care of God's earth. That is my way of changing the world.

CHAPTER 1

DANGER AND OPPORTUNITY

DANGER AND OPPORTUNITY

The night crawlers fascinated me as I walked my fields in the drizzle of a May morning. I began counting. Thirty-six of them to the square yard. Dropping to my knees, I sifted a clump of soil through my fingers, and a five-inch earthworm wriggled on my hand.

Generally, these creatures emerge from the soil under cover of darkness, but when conditions are right and the soil is rich, they

It's not hard to find earthworms in the soil of our fields—
every shovelful has them!

come out in the daytime, too. By their presence in such numbers, they were telling me, loud and clear, that they liked living at Cedar Meadow Farm, which my family has operated for nearly a century here in Lancaster County, Pennsylvania. This was earthworm heaven, and you could say they were thanking me for keeping it that way.

To the east, the sun soon enough would rise high to drive away the mist, and the worms would retreat to continue their feast in the moist and luscious layers below. And after a day's labor, with the sun setting across the Susquehanna River, I retreated from the fields as well, back to our farmhouse, to the embrace of family, and to the aroma of my wife's delicious cooking, grateful to my Maker for the gift of another day of doing what I love.

Farmers live by cycles of suns and seasons, of risings and settings, of plantings and harvests. The best of them respect nature with a reverence of love and fear. Every day, farmers of all stripes—from the major commodity producers of corn and soybeans and wheat, to the vegetable growers like me who serve local and regional markets, right down to the backyard gardeners with a shovel and a packet of peas—scan the horizon to watch for heavy skies.

Nature is mostly beyond our control. Farmers can prepare, certainly, and pray, but we are at its mercy much of the time. We learn to save up for the meager seasons while rejoicing in the good ones. If you work the land, you understand. The clouds could bring gentle showers to nourish your crops. Or they could bring a pounding downpour with hail that strips away your high hopes for the harvest. I've been there. I know.

Look to the horizon now. The wind is shifting. Something big is brewing out there, and I'm not talking about the weather. I'm talking about something else that could either sustain you or destroy you. This front that is sweeping in has the power, within a decade or two, to extinguish dreams—but the farmer who sees it coming, and who prepares, and is willing to try something different, will flourish. His cup will run over.

I write this book as both a warning and an encouragement for my fellow farmers. I write it also for anybody who works with farmers as a consultant and advisor, or who just cares about farming, or about nutritious food, or about being a good steward of the earth. I write it for those who care about the Chesapeake Bay, and Lake Erie, and the Mississippi Delta, and anywhere else where land or water have been at risk. I write it for those who buy the farmer's food, whether that's the supermarket consumer or the multinational conglomerate.

This book is a wake-up call. It's time to get up, stretch, eat a hearty breakfast, and get out to the fields to see which way the wind blows.

The Lake Erie algae bloom is easy to see, and it's not good. Our fields should be green, not the water!

THE POWER TO ENRICH

Here it is, in a nutshell so to speak. This is my message to my agricultural friends who make a living off the land:

You are in danger of becoming obsolete. Major market changes are coming that will force farmers to make difficult decisions. If you adjust to those changes, you can avoid the danger and capitalize on a huge opportunity. If you don't, you're a goner. To preserve your livelihood, you must act now.

Here's why: consumers these days are demanding food that is produced responsibly and sustainably. The trend is unmistakable and accelerating. Many of those folks don't live near a farm, and they don't know a farmer, but they do know what they have heard in the news or read on blogs or wherever they get the scoop on what's happening. They learn about chemical-laden waterways and hormone-infused livestock and denuded forests, and though there's plenty of blame to pass around, somehow it's the farmer who often gets the black eye.

On the other hand, consumers deeply appreciate farmers who, in their estimation, do the right thing. People want good food that's good for their bodies and good for the planet, and they will line up to buy from the farmer who they trust can deliver it. Whether they are right or wrong about it all isn't the point. This is what more and more consumers clearly want—and the first lesson of economics is that when supply is limited and demand is strong, prices rise.

Low market prices hurt farmers just as surely as drought or flood or blight. Many are struggling to eke out a living. No surprise there. They face high production costs and dwindling returns. It's a matter of survival. As they try to keep a business going, they naturally will be looking for ways to trim their expenses to make some kind of profit on what they produce. Otherwise, why bother to produce at all? As much as they love it, they might have to leave it.

So what can you do, other than calling it a day? How can you ensure a profitable future? How do you future-proof your farm so that you can keep doing what you love?

One way that some farmers are getting a better price and reducing their costs is by getting on board with what has become known as the regenerative agriculture movement. Now, that might just sound like fancy words, but it really comes down to farming much as it was done in the old days, in the ways that farmers always knew was best—but now they can do it with the advantages of advanced machinery and twenty-first-century technology that will allow them to get back to the basics on a big scale, producing food profitably that is pleasing to both the palate and to the health-conscious and environmentally concerned consumer.

It comes down to waking up the soil—that is, regenerating it, rejuvenating it, restoring it so that it once again can do what it was designed to do. Modern agriculture often has been unkind to the land. Trying to boost yields from overworked fields, farmers for decades have poured on pesticides and fertilizers, paying a fortune to do so. For whatever short-term benefits they got year to year, the food wasn't tastier or

> It really comes down to farming much as it was done in the old days, in the ways that farmers always knew was best—but now they can do it with the advantages of advanced machinery and twenty-first-century technology.

more nutritious, and the soil suffered. The farming methods messed with its biology, basically putting the soil to sleep. We kept it going artificially, like a patient on a respirator.

The land, though, is resilient. Give it a chance, and it will breathe freely. Once the soil begins to function again, the farmer can set aside some of those chemical crutches. By going back to time-tested farming practices, we all benefit. The farmer is happy to spend less money on fertilizers and pesticides. The consumer is happy to spend more money on the farmer who has reduced or given up on all that stuff. And the night crawlers? Start counting. When they're happy, everyone's happy.

What's old is new. Our founding fathers understood a thing or two about good farming, and they never heard tell of the concoctions that distant generations would dump and spray on their crops. They did know how to enrich and stabilize the soil, keep the weeds and pests at bay, and produce bountiful yields. After each harvest of their cash crop, they grew an interim crop to fortify the soil until the next planting. They called it green manure—a living fertilizer that nourished and protected the ground in both dry and rainy seasons.

Today, we call it a cover crop—and I am on the road regularly, far and wide, spreading the news about a concept that is very old. I am a longtime advocate, as well, of a newer concept that also protects the land, and our waterways, too—and that's planting without plowing. Seeds will grow just fine without upending and disturbing the earth. No-till and low-till methods not only safeguard the ground, but also save the time and expense and compaction of all those trips around the field. For decades now I have been urging my fellow farmers to put away their moldboard plows and cease slicing and dicing the soil.

Cover crops and no-till farming are at the heart of regenerative agriculture. They have the power to enrich both the soil and the farmer who employs those methods. They are the building blocks for the future-proof farm.

No-till planting corn into a nice cover crop mix of hairy vetch and crimson clover. This reduces the need for fertilizer and herbicides.

A CENTURY IN THE RIVER HILLS

Here in southwestern Lancaster County, I work the farm that my grandparents bought in 1935 after they married. The farmhouse was built about 1890, which was around the time that this area was being cleared for agriculture. The locals call this the River Hills region because it is close to the Susquehanna, which flows down to the Chesapeake Bay, and the terrain is hillier and more wooded than the rest of the county. We are off the beaten path of the tourists who come to see the Amish and their ways. It's quiet here.

My father and mother, Elias and Marian Groff, purchased the adjoining property, with a farmhouse of about the same vintage, in 1967, and it was there that I grew up. By the 1980s, when I was a young man in my twenties, my grandparents had passed on. When I married, I moved with my bride, Cheri, to the original homestead, where we reside to this day. Here we raised our three children, Dana, Lauren, and David, all now adults leading active lives.

In all fairness, I should say it is David, not me, who works the farm now, since I have bequeathed to him most of the field labor and some of the management decisions. That makes him the fourth generation to whom this land has given a good and honest living. Our intention is that he will gradually be taking over the farm, just as I took it over from my dad nearly four decades ago. We farm about three hundred acres, two hundred of which we own and the rest we rent. We raise a long and diversified list of crops, including small grains and corn, hay, heirloom tomatoes, pumpkins, squash, oilseed rape, and CBD hemp. We also grow cover crops for seeds.

There wasn't a day when I wanted to be anything but a farmer. When I was in the sixth grade, I informed my mom that there was no point in my going to high school because, after all, farmers didn't need all that stuff from books. She asked me to write down why I felt that way, and I eagerly tried to convince her. I did go on to finish high school, and I'm glad that she nudged me because I had a pretty good time. I didn't apply myself all that much, though. Homework wasn't my thing. I much preferred field work.

No way was I going to college, though. The only degree I wanted was my dad's approval as I worked alongside him. He saw my passion for the farm life, and he was offering me the opportunity to slowly take over the operation. I was learning plenty about farming, about business, and about life. I was learning about economics, and profit margins, and taxes, and investment. Above all, I was learning God's purpose for me. It seemed as if going off to college would just slow down my education.

*I've always wanted to be a farmer! That's me on the right with my sister
Evonne in the center and younger brother on the left (deceased).*

To this day, as I travel and speak around the country and around
the world, people often ask me, "So where did you go to college?"
To which I answer: "I'm still in college." I get some odd looks, but
it's true. Every day I'm learning more, and every day I try to pass
some of that knowledge on to others who I'm convinced will find it
helpful. I've long since lost that piece of paper on which I scribbled
my rationale for wanting to drop out of school. Turns out my mother
was right, though. It was never going to happen. I was destined to be
a student for a lifetime. And a farmer, too.

In my years at Cedar Meadow, I have tried to be a good steward.
My efforts have improved the quality of the soil to the point where the
amount of organic matter has nearly tripled. And word gets around.
The way we grow our tomatoes and squash and other produce has
attracted the attention of people who want to buy it from us. They
are willing to pay a good price for good food—and that's the lesson

that farmers must take to heart as they try a different approach to their marketing.

LEARNING BETTER WAYS

My family has seen a world of change in agriculture in the nearly ninety years since my grandparents purchased these acres during the depths of the Great Depression. In those days, farmers commonly parceled the land up into square fields of more or less equal size, without regard to the slope of the terrain. In the spring, my grandfather would plow his fields, turning the soil over before pulverizing it with a disk and harrow. That's how farmers prepared their fields. They got a smooth planting surface, but they left the soil vulnerable to cloudbursts and erosion and the baking rays of the sun.

If you compare aerial photos of Lancaster County from the 1940s with ones from the 1960s and 1970s, you can see a dramatic change in the landscape. The federal Soil Conservation Service was encouraging contour strip farming, which fundamentally changed agriculture in our county—and on our farm, as well. Instead of plowing a whole hillside, farmers set up their fields in strips that followed the slope of the land, and they didn't plow every strip. They kept every second field in grass, which they could mow for hay. That way, a downpour was less likely to carry the topsoil down the hillside. Farmers still plowed, but this method protected them from a complete washout.

That was the way we farmed when I was a boy, as did the neighbors all around. Then, in 1982, about the time that I was escaping from high school, I took an interest in the no-till method, which can significantly improve soil quality and biology and reduce farming costs. I worked hard to convince my reluctant dad. On several fields close to our house, where the slope is about 15 percent, we gave it a try, and it worked out well—in fact, those fields have never seen a plow since. Gradually, we shifted more fields over to the no-till method.

Though I was hitching up the old moldboard less often, the plowshares still were shiny from turning the earth each planting season. I felt I couldn't give it up altogether. "Hey, Steve, when are you going to sell that thing?" my cousin, who was also a farmer, asked me one day. "I can't," I answered. "I still need it for the vegetables." No-till was just fine for some of the crops, but how was I supposed to plant, say, my tomatoes and pumpkins that way? I felt that I still had to plow to get those in the ground. And that's what I believed until 1994, when I met Aref Abdul-Baki.

Dr. Abdul-Baki was a research plant physiologist at the Department of Agriculture's vegetable laboratory in Beltsville, Maryland, and I attended a presentation he gave at a Pennsylvania Association of Sustainable Agriculture conference. He spoke about his work with no-till transplanted tomatoes planted directly into cover crops. I was impressed. It was as if this man had discovered the missing link. He had been comparing notes with Dr. Ron Morse, a horticulturalist with Virginia Polytechnic Institute. Morse had developed a no-till cabbage planter for the hills of southwestern Virginia, where farmers who still plowed steep hillsides had found themselves mired in mud. Together they realized that they could develop systems for the no-till transplanting of a wide variety of produce. I was already committed to no-till for corn and soybeans and alfalfa and wheat on my farm. If I could put this research into practice, I thought, my farm could be fully no-till. No more hours spent running around in circles, quite literally, dragging the plow.

That summer, I drove down to Maryland to see for myself what Dr. Abdul-Baki was discovering in his research fields. There, in his test plots, I saw healthy tomato plants, flourishing in undisturbed soil. I began working with both those men—if they could do it, so could I—and eventually designed my own version of a two-row no-till vegetable transplanter that functioned better for our farm. Around that time, I also designed the first implement for cover crop rolling in the United States. That was a major innovation for which I am known.

The roller crimper I developed and used since 1995.
This is the first one built in the U.S. I still use it today!

As for my pumpkins, I started tinkering with my no-till corn planter, and after some adjustments it was doing with pumpkin seeds what it had been doing with corn seed. The pumpkins, too, we were able to plant directly into a cover crop. And that was it. I was 100 percent sold on the no-till concept. To this day, since the mid-1990s, none of the fields at Cedar Meadow Farm has been put to the blade. The old moldboard plow grew tarnished in its retirement, and I eventually sold it.

*No-tilled butternut squash planted into a nice rolled cover crop—
very clean, resulting in less disease.*

*Butternut squash with no ground cover.
Notice the soil is splashed all over the leaves.*

MAKING FARMING FUN AGAIN

In the years since, I have been an emissary for soil health, spreading the good word about no-till and cover crops and the huge opportunity that regenerative agriculture offers. This small farmer from southeast Pennsylvania—from Lancaster County, which is considered by many to be the horse-and-buggy epicenter of traditional American farming—has been privileged to articulate principles of agriculture to audiences around the world.

I have been invited to speak by prominent national and international companies, and they keep inviting me back. I have spoken in auditoriums and meeting houses and at farmer field days in most of these United States and in most of Canada's provinces. I have talked to farmers and plant breeders and corporate executives in Europe, Australia, South America, and South Africa. This has been the best kind of whirlwind.

> Farmers listen to other farmers. They are looking for trusting relationships with people who know their stuff, and not just the book stuff.

In my time I have learned that folks are most likely to pay attention when you make it real for them. Some farmers have told me that the internet has been the source of almost everything they have learned about regenerative agriculture. There's a lot out there to soak in, for sure, if you can separate the fact from the fiction. The strongest connections that I have made, though, have been one-on-one. No matter how much you read, there comes a point where you will have some specific questions. On the internet, you might find some answers, and then, just a click or two away, somebody else will be saying just the opposite. Whom do you trust?

And that's my point. Farmers listen to other farmers. They are looking for trusting relationships with people who know their stuff, and not just the book stuff. To make it real, you need to spend some time in the other guy's boots—walking the fields together, sifting the soil, sharing your frustrations and reservations, your ideas and doubts. That's how you connect. Once you develop a relationship, your advice means something, and that farmer will know that you are just a phone call or an email away.

Meeting with farmers on their farms in Argentina.
These are some of the most meaningful times in my international travels!

That's not to imply that farmers are out of touch with the online world, but most of them like a handshake, too. Few are inclined to speak in front of a group of people, particularly a large corporate crowd, but they are listening intently out in the audience. They would rather talk privately in the hallway or at a coffee shop. I am comfortable talking to big gatherings about important matters, but I can do that only because I am able to listen to individuals, too. I'm not there to wow everybody with some fancy roadshow.

I have found my niche as a public speaker and an advocate for a better way. I never expected this when I was that kid arguing with his mom about going to high school, but God had other plans for me. He gave me a gift and a purpose. I am a connector. I draw people out, hear what is on their minds, and offer suggestions. I don't pretend to have all the answers, but I'm always happy to hook people up with others who might know more, whether it's a horticulturalist in Australia or a farmer in South Africa. There's a world of knowledge out there, and my mission is to build bridges.

Over and over, I have heard the same message from those who have made the changes that I advocate: *This makes farming fun again.* Farmers often get discouraged. They have done what the other crop consultants or their co-op have suggested, but the land just doesn't seem to be responding the way it once did. The soil is getting worn down, and the farmers are feeling worn down, too. And then they switch to cover crops, and within a few years they are seeing a big difference. The soil is richer, healthier, more productive. It doesn't need a fortune in chemicals to keep it going. They have seen the land thrive by the work of their own hands, and they feel fulfilled. It's fun.

To get close to the land is why most of us became farmers. I tell my fellow farmers to never stop walking their fields and taking a good look around. As Yogi Berra put it: "You can observe a lot by watching." Get up close and personal. Knowing your soils is the art and the heart of farming, and to practice that art you must sink your fingers into the earth and understand what you should be looking for. The soil will tell you what it needs, if you pay attention.

Up close and personal: showing Dr. Ray Weil's college students how to assess soil health in my fields.

CONTROLLING YOUR DESTINY

The first rumblings of thunder came as I headed out on the tractor to where acres of alfalfa lay ready for baling. Looking up at the darkening sky, I pulled up the throttle and called to Chad, one of the three neighbor boys we hired for summer harvest help.

"Hook up the baler and jump on the wagon," I said, feeling the pelt of a raindrop on my arm. The forecast had called for fair skies, and I hoped the storm would hold off, even for an hour or two, until we got as much of that hay as we could into the barn. It was cured and perfect and would bring a good price—unless it got soaked, in which case half its value or more would wash away, and we'd have to spend time tedding it and raking it again to dry.

The tractor and baler roared as I guided the machinery around the windrows while the neighbor kids stacked the bales six high on the wagon, and then it was back to the barn to unload. I left two of the boys busy with that chore while I took Chad out to the field again to continue the race against time.

"I've got to move fast," I told the sweating teenager as we hitched another wagon to the baler. "I need this money. I've got bills to pay."

Soon the skies cleared—it was a false alarm, with only a brief sprinkle, as I recall—and we got plenty of sweet, dry alfalfa into the haymow. It didn't always work out that well, though. Years later, Chad told me that he had never forgotten my words that day. He could see what was at stake, and it was nothing short of the farmer's livelihood.

People are fond of saying that life on the farm is kind of laid-back, but any farmer will tell you that it is kind of stressed-out, too. It's hard to understand that yo-yo feeling unless you grow up with it. I recall a hailstorm long ago, back when I was still working with my dad, that shredded our tomatoes just as they were in their prime. We lost thousands of dollars from just that one storm—several acres of red beauties, left cut and bleeding. Things happen. You don't blame God. You do your best, minimize whatever risks you can, and keep on keeping on.

The reality is, sometimes something completely beyond your control can slow you way down, or even stop you in your tracks. The dairy industry, for example, is in crisis. That's no exaggeration. Like every industry, it has had its ups and downs, but never before in my lifetime have I seen so many farmers selling off their cows here in Lancaster County. Milk prices are low, consumers are turning to alternatives, and a turnaround doesn't seem likely anytime soon. The dairymen have little control over their market—nor do the corn and soybean growers, for that matter. Farmers in all sectors of agriculture are vulnerable to worldwide influences with the power to determine their fortunes.

One of those factors is the change in public tastes—and what is at stake, if you are farming, is surely nothing short of your livelihood. This time, not only can you minimize the risk, but you can capitalize on the opportunity. The big changes that are coming down the pike can either make you or break you. Cover crops are more than a good thing for soil health and the environment. They will be essential for farmers' financial health as well. As the marketing of farm products changes dramatically, those who fail to adjust to what is in demand could be left behind and end up with no market at all. Every farmer should be saying what I said out in the alfalfa field that day: "I've got to move fast. I need this money. I've got bills to pay."

The market can be as capricious as the weather. Put those two together and you can see why a lot of farmers have been gritting their teeth—but the good news is that cover crops and no-till farming offer a measure of control over both those variables. Aside from hail, crops fail when conditions are either too dry or too wet, and these practices result in a more resilient soil that improves the yield no matter what the climate has in store. In other words, both the soil and the farmer are better able to deal with the weather. And because consumers these days buy from farmers whom they believe produce healthier food, the farmer will be better able to deal with the market.

This is how farmers can assert control over their own destinies. They need not despair over the market. Instead, they should anticipate it. Any good business, small or large, studies the market, takes surveys, and tests the waters to identify consumer trends, looking for opportunities to tap into a growing demand. Such companies are looking to join forces with the progressive farmer who is willing and able to satisfy the consumer. If you answer their call, you could thrive. If you ignore them, they will ignore you. The market, in short, is what you make of it.

THE MOST PRECIOUS OF ASSETS

Here in the Chesapeake watershed, a lot of farmers have long felt defensive about their role in the pollution that damaged the ecology and the economy of the bay. For decades, agricultural runoff of sediments rich in nitrogen and phosphorus and laden with contaminants have contributed to the depletion of the fish, oysters, and crabs that for generations supported so many families. *We don't like pollution either,* farmers responded, *but our job is to feed people. If folks want to eat, we need to farm this way. It's just how it is.*

That mindset has shifted significantly. Along with the negative publicity has come widespread education about the issues, and farmers are not against learning more so long as it also makes good old-fashioned common sense. They began to realize that keeping the topsoil on their fields was keeping money in their pockets. After all, they paid for that soil when they purchased the farm, so why should they be letting it slip downstream? Better to do whatever they could to hold that soil in place and unlock the natural nutrients within it to produce healthier crops. Farmers who do well are fundamentally good businessmen and good managers who know how to protect their assets—and the soil beneath their feet is the most precious of those assets.

That is why more farmers have been exploring the no-till and cover crop concepts and how to best adapt those practices to their land. They certainly are not out to harm the very thing that sustains them. Once they have invested in soil health for the long term, they will gain an edge over the competition. They can see that maintaining healthy soil is not only a win for the environment, but also a win for their bottom line. It makes good farmers better.

Major corporations—General Mills, Wrangler, McDonald's, and Patagonia, to name just a few—are paying close attention. Their customers care, so they care, too. If the big guys are investing in cover crops and no-till production, you can be sure that they have their reasons. They see how the market is evolving, and they are preparing. They are positioning themselves for success.

Recently, I met in Europe with representatives of the French-based company Bonduelle, the world's largest frozen vegetable processor, to consult with them and their growers about the use of cover crops and no-till methods, specifically for sweet corn. The company has taken a keen interest in making sure that the food it sells globally is produced responsibly. Its motto, in fact, is *La nature, notre futur* (Nature, our future).

And it is not the only company of such magnitude that is taking an interest. As I was speaking at a recent conference on cover crops in Indiana, I looked out at the audience and saw, seated near the front, representatives from Archer Daniels Midland, the largest buyer of corn and soybeans in the United States. And just two days before that, Cargill was represented at a soil-health conference where I was speaking in Kansas.

These are some of the world's largest food companies, and they don't fly their people to conferences around the globe just to enjoy hors d'oeuvres. They are listening intently to what I have to say because they are all but certain that the regenerative farming movement will translate into sizable profits at some point.

This is what I am telling farmers, wherever they, too, will listen: think like the big guys. Follow the market. It takes a lot of effort, but farmers have always had a penchant for hard work and ingenuity. This is your chance to really shine at something new. In a way, though, I'm saying just keep doing what you have always done, because you already know what works. Wake up early, check the weather, and tend to your fields. It's a new day. Get a good start.

CHAPTER 2
SHE'S YOUR BOSS

SHE'S YOUR BOSS

A young woman clips through the aisles at the supermarket, filling her cart with the week's groceries as her toddler, who is seated in front of her, babbles while bouncing a colorful box of cereal. This shopper is decisive and efficient, typical of her millennial generation, but she is in no hurry when it comes to examining the quality of what she is purchasing. She reads the labels. She considers the nutritional content. In the produce section, she looks for food that has been grown responsibly and sustainably, and she is willing to pay more if she believes it comes from farmers who care.

"Meet your new boss," I tell fellow farmers as I speak to audiences about the virtues of cover crops and other environmentally sound agricultural practices. "Whether you like it or not, your future is in her hands."

I borrowed the "meet your boss" idea from a series of reports published back in February 2016 in *Successful Farming*, one of the foremost agricultural journals in the United States. The magazine looked at consumer trends involving the millennials, trends that have continued to strengthen in the years since. *Farm Journal*, another leading agriculture magazine, launched "Trust in Food" a few years ago to educate farmers and consumers on what the agriculture industry is doing to build consumer confidence and trust in the way our food is grown.

The millennials are a tech-savvy bunch, generally, who aspire to make the world a better place, and whose penchant for idealism extends to what's going on down on the farm. All in all, they care deeply about what they feed their families and what has been happening to the environment. Their focus increasingly has been on good taste, good nutrition, and good practices. They favor simple, fresh, unprocessed, authentic foods—and as they pursue healthier lifestyles, they are raising their kids to do likewise. That's not to say that millennials invented the idea of eating better and taking care of the earth. In their desire to make a difference, food-conscious consumers of any generation want to partner with farmers who want to make a difference, too.

Until a few generations ago, though, consumers didn't much question how their food was produced. Backyard gardens were more common back then. Families didn't doubt that what they proudly had grown themselves was wholesome. After all, they had tended to it with their own hands, planting it, nurturing it, harvesting it. They presumed that farmers likewise would deliver wholesome foods, produced with pride, to the local market. For decades, the food consumer simply trusted that all was well.

Eventually, most folks came to understand a sad truth. As small family farms gave way to developments, those that remained relied increasingly on a variety of chemical potions to improve yields. So did the mega-farmers with thousands of acres under their dominion. Although those yields went far toward feeding a hungry planet, they came with environmental consequences that gained widespread publicity. Public sentiments shifted sharply, particularly after the publication in 1962 of Rachel Carson's *Silent Spring*, an influential indictment of the widespread use of pesticides. Agriculture took a lot of the blame for pollution.

Today, in the age of social media, many more people are talking about how their food is produced and questioning whether Big Agriculture has continued to chase the big bucks at the expense of their health and the planet's welfare. They would prefer to somehow return to a day when farming felt friendlier, more neighborly, when the producers wore bib overalls out in the field, not a business suit in the boardroom. People still cherish the enduring image of the farmer as devoted gardener.

> People still cherish the enduring image of the farmer as devoted gardener.

It seems to be human nature to take for granted whatever we have in abundance. In much of the world, families must spend a third or more of their income on food. A household in Nigeria spends more than half. In the United States, food remains relatively cheap, at only 6 to 7 percent of the family budget.[1] With food bountiful and inexpensive, Americans long paid little attention to how it was grown. However, the prospect of anything threatening our breadbasket is bound to put a spotlight on agricultural practices. The question becomes: Is the American farmer treating the land like the precious resource that it is? Are we being good stewards of our blessing?

As environmental and social awareness continues to take root in our society, many Americans sincerely want to do their part. They have been willing to spend more for food if they are convinced that it was grown responsibly by farmers they respect. As consumers take notice, the marketers follow suit—and so the trend accelerates.

---

1   Alex Gray, "Which countries spend the most on food? This map will show you," World Economic Forum, December 7, 2016, https://www.weforum.org/agenda/2018/12/this-map-shows-how-much-each-country-spends-on-food/.

If you make your living from the land and hope to keep doing so, then listen carefully to what the consumer is saying. Unfortunately, many farmers are only slowly beginning to understand the power of that discerning shopper filling the cart at the market. She can make them or break them. When I describe her during my speeches, the conference room tends to get very quiet. It is a scenario that many in the audience might find disconcerting. But everyone knows it's true.

"I'm not saying she's always right," I quickly add. "She doesn't know more about farming than you do, by any means, and what she thinks is the truth isn't necessarily so. She probably hasn't spent any time out on a tractor or tried to get by on a farmer's paycheck. That lady likely has no clue what goes into running a farm." She might not consider, for example, that those organic berries perhaps were trucked for thousands of miles, leaving a carbon footprint from coast to coast. She might not realize that those happy free-range chickens pictured on the carton don't look much like the ones that produce eggs at a price that she can afford.

When they hear me acknowledge that the shopper could be wrong, a lot of the farmers nod in agreement. It's as if they want to shout, "Right on, Steve, you preach it!" They seem to feel better putting her in her place. *We know what we're doing,* they're thinking. *We've always farmed this way, and who is she to tell us our stuff?* Not everyone in the audience sees it that way—but I know that some of them do.

"The thing is," I continue, "whether she's right or wrong isn't the point. What matters is what she thinks is right, because that's where her money goes. That's how she's filling her cart. She's the one who decides which farmers will stay in business. Don't think you can ignore the boss—because, well, the boss can fire you."

That supermarket shopper may soon be equipped with instant

data to back up her assumptions. We are on the verge of technological advances that will give consumers unprecedented abilities to assess their food before they buy it. Imagine this scenario: the shopper comes to the market equipped with a device that she can point and click at a tomato, or a stalk of celery, or an apple. She might even have downloaded an app that lets her do that with her mobile phone. She gets a readout of the levels of twenty-five minerals and nutrients—and the prize goes to the farmer with the highest score.

I recently was one of three hundred people who participated in the beta testing of a nutrient density meter under development. The meter connected to my laptop with a USB cord. And it wasn't just for vegetables. I used it to analyze milk and other products. The technology is here. Consumers won't be shelling out the big bucks that such a unit would cost now, but prices tend to drop as technology advances. What will be the implication for farmers when consumers can easily afford a nutrient reader and take it along with them to shop for groceries? Or when the co-op tests a load of corn and sets the price not by the quantity but by the quality of its nutrients?

That time is coming, and it will be a game changer in agriculture. We're not talking science fiction here. Farmers must prepare for the day not far away when the buyers, whether at the grocery store or at the grain elevator, will hold that power in their hands.

I tell my listeners that they cannot win a battle of perceptions. If they tried to engage the typical supermarket shopper in a debate about, say, whether genetically modified foods are good or bad, or whether a product labeled organic is all it's cracked up to be, they would just be going down a rat hole. Once people have made up their minds, they look for whatever research will back them up. The big question is: who's paying for that research? Who's disseminating that data? I know the talking points on both sides. I could argue them either way, if I chose to do so. I don't. Controversies polarize people.

"What we really need to do is tell our story," I say. "That shopper isn't the only one who cares about responsible farming. Every one of you cares, too, and she needs to understand that we are on her side. That's why you can't go wrong with cover crops. She's not going to argue against something that protects and enriches the soil, that stops its nutrients from flushing away downriver, and that keeps pollution out of our waterways and nitrates out of her drinking water."

There's no downside, I explain. It's an offer she cannot refuse. Anyone who appreciates nutritious and delicious food, grown with pride and respect for the land, will want more of a good thing when they find it. "Once she learns what you're doing with cover crops," I tell the farmers, "the boss will see you as a hero."

THE HEROES OF HEMP

In 2019 I embarked on a new venture for Cedar Meadow Farm. It is one that I am confident tells the story that consumers need to hear about responsible, regenerative farming that pays attention to a changing market. In the first year that it became legal in Pennsylvania, I planted seventy acres of CBD hemp—along with several species of cover crops—without tilling the soil.

Let me be exceedingly clear: this is not marijuana, and I am not promoting marijuana use. What I am growing cannot possibly get anyone high. CBD is the chemical compound that can be extracted from this variety of hemp for a range of proven medicinal purposes. Although it is from the cannabis family, this is not the variety with the THC compound that is psychoactive. In fact, I must strictly test my crop to demonstrate that any trace of that compound is so low as to be virtually undetectable.

One of my CBD no-tilled hemp fields.

Think of it this way: farmers grow various types of corn for different markets, such as sweet corn for the dinner plate, field corn for livestock, or popcorn for, well, watching movies. I once heard a story about some city tourists driving down a country road in Lancaster County, eager to see buggies and such. As they passed field after field of tall, green corn, they figured nobody would miss a dozen ears and they could save five bucks at the market. They were in for a surprise at dinnertime that evening, and it wasn't a sweet one. They had snatched field corn intended for cows. They just didn't know—because, after all, the leaves looked much the same, and they figured that corn was corn.

That's the way it is with hemp. Industrial hemp has long been grown to make products such as rope, textiles, and bioplastics. The hemp that I grow produces the compound with pharmaceutical applications. It's all in the cannabis family, but people just don't understand the differences. That's why CBD hemp was banned for so long:

the government lumped it in with marijuana even though it serves an entirely different and highly beneficial purpose. The leaves do look much the same, though. Lest anyone get any ideas, I place signs along my fields alerting anyone who should happen by that they haven't stumbled upon a marijuana operation. I don't want any tourists or local teenagers figuring that nobody would miss a plant or two.

What differentiates my hemp crop is how I grow it. Typically, farmers till a field and cover it with plastic to keep the weeds down. Then they will plant hemp transplants through the plastic. I don't plow, and I don't use plastic. I remain true to my conviction that no-till planting and cover crops keep the soil in prime condition.

As one of the top farmers pioneering CBD hemp production in Pennsylvania, I am proud to say that I am growing it to the expectations of today's consumers, using the farming methods that they clearly have communicated are important to them. I have added to my farm a crop that they clearly have communicated is valuable to them.

Dozens of other farmers in Lancaster County also planted CBD hemp during the first year that it was legal here, including a lot of Amish dairymen who set aside a few acres to help them get by while struggling with dwindling milk prices. Across Pennsylvania, the new hemp growers numbered in the hundreds. They were paying attention, and they saw the possibilities.

What I saw, in addition, was a new way that I could influence agriculture for the better as a consultant, advising farmers on how to adapt to a changing market and further future-proof their operations. To that end, I established a network for field training in successful CBD hemp production with the goal of continuing to spread the word that no-till and cover crops are still the way to go. These will be the farmers who do it right. They will become, in the eyes of their caring customers, the heroes of hemp.

Cover crops surrounding my no-tilled CBD hemp.

FOLLOW THE LEADERS

Whenever more people are willing to pay a premium for a product, you can be sure that businesses will snap to attention to tap into that demand. Entrepreneurs are interested in what interests their potential customers. As that demand rises, so will the price of the product—and so will the entrepreneur's potential profit.

Nothing wrong with that. Our economy thrives on the profit motive. That's why things get made and done in our free society. Consider, for example, the cost of fair-trade coffee in a specialty shop. It's a price that people are willing to pay for the feeling that they are doing what's right. If folks didn't want it, those shops wouldn't be selling it. Demand encourages merchants to rise to the occasion and figure out how to serve the customer better than anyone else does. When our system is operating properly, the price of a product sets itself at a fair level that reflects availability and desirability.

When it comes to the growing demand for food that is produced responsibly, it is not just niche businesses that are seeing the light. The big corporations and their marketers are paying rapt attention to what goes into the carts. Striving for the competitive edge, they value consumer insights on how to do their job. Those corporate thinkers see what is coming, and they are getting ready. It's a matter of good timing.

In the mid-1990s, some of those big players were positioning themselves to cater to the "locally grown" movement, which was then in its infancy. The definition of *local* was getting stretched to include food shipped great distances from other states. That's when I met Michael Rozyne, who in 1996 founded a nonprofit food distribution business near Boston called Red Tomato. His goal was to connect truly local farmers with supermarkets in the New England area. He and I later worked out a deal with Red Tomato to market my tomatoes to a leading supermarket chain, and I appreciated his attention to detail and to ensuring the farmer got a fair price.

Today, the big players remain alert to new trends. Consider that General Mills, as of 2019, had put about $4 million into soil-health research and promotion[2] and has been sponsoring regenerative agricultural conferences. That's just one example. As we will see, a variety of other major players have been keeping in tune with their customers and investing in what really amounts to just good farming. Even big-box stores such as Walmart are stocking up to give people what they want. It's just smart business.

As they analyze the demographics and psychographics of who might buy their wares, the marketers confirm the vibe of the buying

---

2   "General Mills to advance regenerative agriculture practices on one million acres of farmland by 2030," General Mills news release, March 4, 2019, https://www.generalmills.com/en/News/NewsReleases/Library/2019/March/Regen-Ag.

public: people want greater transparency all along the food production chain. The social media have accelerated that trend, fostering communication and greater awareness of environmental and health issues. Consumers are tuned in and looking for agricultural practices that sustain and regenerate.

We supply trailer loads for butternut, spaghetti, and acorn squash for supermarkets in the mid-Atlantic region.

Farmers want to be good caretakers, too, which means their interests jibe with what the public desires. It is time now to listen closely. A farm operation is a business, and astute businesses are ever alert to whatever might influence their market. Smart farmers will be getting ready to seize the opportunity and serve this new market rolling in.

It's a matter of good timing for the farmer just as it is for the corporate executives. A summer rain, for example, nourishes the crops and replenishes the soil—but if it comes when you have acres of alfalfa lying cured for the baler, you just might feel like shaking your fist at the sky. Let me assure you: that's not going to do much to change your situation. Those showers might come at an inopportune time, but you know full well that they are essential for your farm's survival and growth. You shouldn't be out there cursing a blessing.

Today's future-proof farmer adapts quickly, turning responsible farming into a lucrative business. If the major players are positioning themselves, shouldn't you, too? This is no passing fad. This is a prime opportunity that must not be missed. It's time to follow the leaders. As a vegetable grower on the front lines of the issue and a longtime advocate of cover crops, I have a unique vantage point. My aim is to alert you to what's coming so that you can prepare.

> Today's future-proof farmer adapts quickly, turning responsible farming into a lucrative business.

You will find a booming and broad market for your new endeavors. Not only will the millennials be eager to buy from you, but so will their parents and their children. Though some consumers will always shop for a bargain and look for the rock-bottom price before anything else, many others will gladly pay extra to support a cause, or to feel included in a movement, or to chalk up an experience. That's the nature of millennials.

That supermarket shopper aiming her smartphone at the produce will be looking for more than just the levels of nutrients. She also will want to hear a good story that makes her proud of her food choices. Cover crop farmers can give her that story, and it is a good one indeed—if we choose to tell it.

CHAPTER 3

THE POWER OF PUBLIC PERCEPTION

THE POWER OF PUBLIC PERCEPTION

Often, it takes only a few words for me to sum up what I do and why it matters so much. A single sentence can transform me in people's eyes from an ordinary farmer into a hero of sorts. It's not as if I'm swooping in with a cape, but I kind of like the feeling that folks appreciate my mission.

Whenever I encounter someone who clearly doesn't quite get what cover crops are all about, I can enlighten that person on the spot by saying something like this: "A cover crop is what helps to keep nitrates and sediments out of the Chesapeake Bay."

That's how I explain it here in my neck of Penn's Woods, upstream from the bay. Depending on where I'm visiting, I might say that a cover crop is what keeps the bad stuff out of Lake Erie, or out of the Mississippi. I pretty much can just plug in the name of the regional watershed.

Wherever I go in the world, it seems that the locals have been hearing for years, often decades, about some nearby environmental nightmare. By putting cover crops in that context, I get an immediate nod of recognition. If I'm speaking to an audience, I can feel the mood of the room change. I have captured everyone's attention and know they are listening intently. That all goes to show three things are widespread: the problem, the publicity, and people who want to do something to clean up our act.

"That's great, Steve!" my listeners typically tell me once they see that I've come to town to help save the bay. Or the river. Or the lake. Or their backyard wells, for that matter. For years, they have gotten an earful about what has been washing downstream with every downpour or leaching through the soil to mess up the water that they drink, and that the livestock drinks, all the way to where it can kill the fish or oysters or crabs.

The details differ from place to place. Australia, for example, gets little rain in much of the country, so water conservation is the focus there. The Aussies aren't so concerned about nitrates and chemicals and sediments washing downstream from cropland. Instead they want a way to help catch and keep every possible drop that falls from the sky. What impresses them is the ability of a cover crop, once it is established and appropriately managed, to retain moisture. They care more about keeping the water from leaving than they do about where it might be going.

It's a matter of perception. Cover crops can work their magic anywhere, but why they are considered most useful depends upon the climate. When I visit an area with, say, thirty or more inches of rain a year, I emphasize how well cover crops can control storm runoff, keeping the soil from washing away and carrying nutrients into the waterways, where they cease to be a good thing. When I visit a relatively dry area, I emphasize how well cover crops can retain the moisture in the soil.

That being said, drier areas require a higher level of cover crop management. There are times when it's so dry that a cover crop would not germinate. Or a cover crop could take away moisture that a subsequent cash crop would need to germinate. However, there are times when arid regions do get adequate rainfall, and that's when a farmer should be ready to take advantage of that opportunity. When a few good cover crops are added into the cycle of cash crops, the soil soon becomes better able to hold moisture.

Whatever the level of precipitation, everyone can appreciate that a cover crop contributes to good soil management by adding organic material and natural nutrients that the cash crops need to thrive. Once farmers have established cover crops, along with no-till methods, they eventually see the results: the root systems, and the earthworms that feed on the rich remnants, have loosened and aerated the soil so that it can soak in a lot more water that otherwise would have been lost to runoff or sucked up by the sun. We've been beating our soils to death for decades, so the recovery will take time. You could say we're paying for the sins of the past, but the reality is that farmers were doing what they thought best. As we learn from the past, we can patiently rebuild our soils to provide nutritious food into the future.

In talking with most of my fellow farmers about cover crops, I can quickly get to the finer points such as those. They are familiar with the concept and often are well-versed in the details. Not everyone, though, has heard the message, and that's okay. My job is to come alongside people wherever they might be and get the conversation going. I find out how much they understand, determine their views, and help them to fill in the blanks. "Have you seen those green fields in December around these parts?" I ask folks who don't know much about farming. "Good chance you're looking at a cover crop."

Here, then, is the least you need to know: farmers plant cover crops to grow during the off-season of their cash crops for the purpose of protecting the soil, enriching it with less dependence on chemicals, keeping it in place, and controlling weeds and pests more naturally. Generally, farmers don't grow a cover crop for profit, although they might. To learn more of the basics, see the brief section before Chapter 1 called *What is a Cover Crop?* It will give you the foundation for building a prosperous future.

SOMETHING OLD, SOMETHING NEW

Once I explain to people how cover crops benefit the environment and can help farmers earn more for their hard work, they often make this observation: "Cover crops make so much sense that you'd think a lot more farmers would plant them."

"Yeah, you'd think," I respond. And plenty of them do, of course, but I run into other farmers who haven't given much thought to cover crops at all. They seem set on holding on to the past. I agree that we are often better off with the tried-and-true—but trying something new shouldn't be out of the question. There's nothing evil about progress, so long as we are bettering ourselves and our world.

It's not as if cover crops are anything new, anyway, though they weren't called that until the late 1900s. Farmers have been planting them for ages—which demonstrates, as the Good Book says, that there's nothing new under the sun. Lucius Columella, a Roman legionnaire turned farmer and agricultural authority, recommended black medic, vetch, and lupine, which "affords an excellent fertilizer for worn-out vineyards and ploughlands."[3] He lambasted landowners who didn't seem to care how they treated the soil, a criticism no doubt familiar to farmers today.[4]

My family and I recently visited Mount Vernon in Virginia, and for us it was more like a pilgrimage, such is our interest in history and

---

3   Lucius Junius Moderatus Columella, *On Agriculture*, Trans. Harrison Boyd Ash (Cambridge : Harvard University Press, 1960), Book I, 157.

4   "Using Cover Crops: A History of Green Manure," University of Minnesota Extension, accessed May 11, 2020, https://local.extension.umn.edu/local/article/using-cover-crops-history-of-green-manure.

our shared reverence for George Washington and all he accomplished in the birthing of our nation. He was at heart a farmer, as was Thomas Jefferson. Our founding fathers differentiated between crops for sale and crops for the soil. They experimented with planting grasses and legumes in rotation with barley, tobacco, wheat, and corn.

Cover crops remained a part of good agricultural practice until they fell by the wayside in the mid-1900s with the development of nitrogen fertilizers, herbicides, and pesticides. Farmers came to see those innovations as a more economical and convenient way to enrich the soil, improve yields, and control weeds and insects. That may have been a shortsighted truth, but it was farsighted folly. Time had tested the old ways, and they were still sound.

The concept of cover crops is quite simple, frankly. What can get complicated is learning to use them to their best advantage to reap the benefits. If you're serious about taking this step, you need to be serious about learning to do it right. It will take some effort. A doctor might be the best general practitioner in town but wouldn't try to perform surgery without attaining another level of education and experience.

You don't just magically become a cover crop farmer when you buy a bag of cover crop seeds. It takes work. Success requires knowledge. Just as you had to learn to manage your cash crops, you must learn to manage your cover crops. That means knowing which mixes to use, when to plant, and how long to keep it growing. If a dry spell comes in the early years, for example, you might have to terminate the cover crop sooner than anticipated so that it doesn't wick up too much moisture. Even when the results don't come the first year or two, you need the resolve to stick with it. After all, not every year is great for your cash crops, either.

Getting everything in balance can be challenging, but that's the nature of farming. Anyone with the backbone to make a living from the land is accustomed to the daily give-and-take and the uncertainty of the seasons. Farmers work hard, pretty much by definition, so it's not a lack of drive, nor downright stubbornness, that holds some back from getting started in cover crops. They don't deny the overall advantages. Still, many remain reluctant. They decline to take part in what should be an obvious opportunity.

The reason is generally economic. A lot of small farmers are living on the edge, with a razor-thin profit margin. Consider the dairy industry: most can't make a living anymore from fifty cows, not with all the competition from soy and almond "milk" and so many other pressures on price. When you're concerned about holding on to what you've got, trying something new feels risky, and just because something is good doesn't mean you can afford it. A lot of farmers simply decide to play it safe—or, rather, they keep doing what feels safe because it is familiar and has gotten them this far. That's the sad reality. That is why more farmers aren't planting cover crops.

> # The public has decided that cover crops make quite a bit of sense, here and now.

You can't reap the dividends unless you make the investment. Those farmers might fear losing ground if they stray from the tried-and-true and try the new, which really is the old. Over the longer term, though, they could stand to lose it all when they find that their market has eroded and no one wants their wares any longer. What holds them back are their own worries, accurate or not, that cover crops might not make sense for them just yet. Maybe tomorrow. Meanwhile, the public has decided that cover crops make quite a bit of sense, here and now. Those buyers aren't waiting on the sidelines. Even now, they are deciding the farmer's future. Such is the power of public perception.

AN EVOLUTION OF ATTITUDES

My farm is a dozen miles south of the tourist corridor on which folks come to see the Lancaster County sights throughout the seasons. As they unload from cars and buses at roadside stands and at quilt and furniture shops, they seek a sampling of the local culture. They come to glimpse the bucolic way of life of a bygone era—the buggies and sleighs and horse-drawn implements of the plain people who abide by the biblical prescript to be *in* the world but not *of* the world.

The tourists come, too, to see the gently rolling landscapes and expanses of fertile fields, often without a power pole in sight. Here, the Amish and the "English," or non-Amish, live and work side by side. To the visitors, it's all so quaint. To those of us who live here, it's all so normal.

In less than an hour, you could walk cross-country from Cedar Meadow Farm westward to the Susquehanna River. That's just a couple miles, and from there the water flows about thirteen miles southeast to the mouth of the Chesapeake, across the Mason–Dixon line in Maryland.

That puts me smack in the middle of the Bay's watershed. The replenishing rain that feeds my fields, and all my neighbors' fields, eventually works its way to the local streams and ever onward to the ocean. The water carries with it whatever it picks up along the way. I guess that could put me among the ranks of those farmers who have been accused of contributing to polluting the bay—except that for years I have been on the forefront, along with many other dedicated souls, of efforts to do something about it.

I'm proud to say that Lancaster County has another distinction besides being an Amish epicenter. This one might not draw more tourists, but it certainly can help to preserve the rural ambience that they come to see. Our local farmers have risen to the challenge and planted cover crops on about 65 percent of the agricultural land in our region. That is among the highest percentages of cover crop adoption in America. About the same amount of cropland here is planted without plows as more farmers have been turning to no-till—and those are conservative estimates.

What has motivated that resolve among the farmers in our community? I don't know a better way to find out than to ask them, and it turns out we pretty much agree on the reason: it was a response to the Chesapeake Bay pollution issue.

Since the 1970s, the Bay's distress has been the focus of public attention. It was found to have an "aquatic dead zone"—among the first ever identified—that was so depleted of oxygen that it could not support life. Fish were dying in droves. The pollution originated from a variety of sources, one of which was agricultural runoff of sediments and nutrients, particularly nitrogen and phosphorous. Farmers took a lot of heat for choking the Chesapeake, and they could see what was coming down the pike. The regulators were on their way.

In those early days, the farmers tended to comply grudgingly. They met governmental requirements to manage manure when they set up "CAFOs," or concentrated animal feeding operations. They submitted to soil tests. They cut back on fertilizer, and they reduced or eliminated tillage to keep sediment out of the Bay. They did what they had to do, even though they didn't particularly appreciate being told what to do by the environmentalists, bureaucrats, and politicians. *We're the experts here,* they maintained. *We're the ones who know how to farm.*

True. And as professionals, they also recognized that they could always learn more and do better, and they wanted to do their part to save the Bay. They understood the importance of joining in a concerted effort. What worried them, understandably, were the costs involved. When you're barely staying afloat, any extra weight could sink you. Their resistance to regulations was primarily economic, just as that would be a major concern as they considered venturing into cover crops as a solution. Those who did, however, experienced less runoff, richer soil, and improved yields—and the word spread.

By the turn of the century, we were seeing an evolution of attitudes. Cover crops and no-till were simply a matter of good farming, and good farming pays. It enhances income, protects the environment, and keeps the regulators away. So many of my friends and neighbors in Lancaster County have followed the drumbeat to a better way. And so many more, I'm sure, will join us, here and in every corner of the world. Good sense has a way of prevailing.

BROADENING THE CONVERSATION

At the edges of farm fields along Pennsylvania's byways, you well might see signs that proudly proclaim, "Cover crops for cleaner water and healthy soil." To spread the word, the Pennsylvania No-Till Alliance distributes those signs to its members during field days to place along land where they have planted cover crops. The alliance, of which I am a board member, concerns itself as much with cover crops as it does with no-till farming. Its slogan: "Farmers improving soil health."

The Pennsylvania No-Till Alliance made these signs to identify cover crops.

The "cleaner water" part of the signs is geared toward the nonfarming public. After all, who doesn't want cleaner water? The "healthy soil" part is geared toward farmers, who want to enrich the soil in their fields and produce better yields while spending much less on chemical fertilizers, herbicides, and pesticides.

Those signs are just one of the ways in which cover crop advocates are actively working toward influencing perceptions—and not just the public's perceptions, but also those within the farming community. Farmers themselves need to understand why cover crops are essential in so many ways—including to their livelihoods.

It's a form of marketing: the alliance is striving to sell everyone on the concept. Farmers pay attention to what other farmers are doing. Like all good business professionals, they want to keep up with the competition. When they see what is working for others, they are more likely to try it out themselves. Enthusiasm breeds enthusiasm. At the same time, a growing contingent of consumers will vote with their dollars in favor of producers who take a stand for healthy soil and cleaner water. It's a winning combination.

There has long been a disconnect between the farmer and the "end user"—that is, the shopper who buys the produce in the market. It goes both ways: for years, the shopper didn't pay much attention to the farmer, and the farmer didn't pay much attention to the shopper. Even today, the closest contact that most people have to a farmer is what they purchase in a grocery store. Meanwhile, the way our system has evolved, farmers have come to depend on someone else to peddle their products to those end users.

Farmers might feel as a result that they grow their crops for the co-op, but ultimately they grow them for the consumer. People will buy what they buy *into*—and they have been handing their money over to farmers who they believe care about much more than money. Agricultural marketing is naturally focusing on what sells—and on the farmers who produce what sells. This is the reality. It's a truth that you will find repeated

> People will buy what they buy *into*—and they have been handing their money over to farmers who they believe care about much more than money.

throughout this book, such is its significance. Farmers who accept it can prosper. Farmers who ignore it will be left behind.

The good news is that fewer folks are ignoring these matters anymore. Farmers and consumers have moved closer to a mutual understanding, and we can thank social media for contributing to this awakening. On their mobile phones and laptops, consumers can vent all their frustrations, if they wish, about the questionable things that they believe farmers do, but likewise the farmers can share with the public, and with one another, all they have been accomplishing with cover crops and other regenerative farming practices. It's a two-way street. Each gets an opportunity to learn more about the other's viewpoint.

All in all, this broader conversation at every level is healthy. Our society benefits when we can talk openly about how our food is produced. That is how we can reach a deeper level of understanding. That is how we repair false perceptions and move from finger-pointing to cooperation.

Farmers have gotten a bad rap. So much negative publicity has tarnished the image of agriculture. With cover crops, however, we have an argument that is ironclad. They are wholly positive and aligned with a strong and growing public demand. The message is clear and powerful. Better farming profits us all.

CHAPTER 4

DO COVER CROPS PAY?

DO COVER CROPS PAY?

I n Chapter 2, we met the millennial supermarket shopper as she filled her cart with produce that she believed was grown by farmers who care about the planet. Let me introduce you now to another kind of shopper, one who clicks through web pages to fill a digital shopping cart. This consumer, too, will have much to say about which farmers will flourish and which ones are likely to fade away.

Deciding that he needs a new pair of jeans, this online shopper begins comparing brands. Landing on the Wrangler site, he notices these words under a video of a modern cotton harvest: "Tough denim. Gentle footprint." He pauses to read: "Every piece of clothing we make, every decision we weigh, starts with respect—both for the planet we love, and for the people who call it home." He browses through pages featuring the "Rooted Collection" of designs celebrating the American cotton farmer's dedication to soil health. He can even choose which of five states produces the cotton for his new jeans and read about the farmers there who "grew" them.

The environmental focus is enough to tip the scale for this shopper, who long had the impression that cotton producers simply wear out the soil, drench it with chemicals, and move on. With a few more clicks, he places an order for delivery to his doorstep. He is part of the wave of countless consumers who have increasingly turned to the internet marketplace to buy a vast array of products, many of which originate from the soil.

The Wrangler example comes to mind because of the key role the company played in a transformation of my approach to advocating the use of cover crops. In July 2017, I was a speaker at a Soil Health Institute conference in St. Louis, Missouri. On the program, I noticed another speaker listed to give a talk—a fellow named Roian Atwood, a representative of Wrangler, Lee, and other brands of workwear. I figured he would probably be giving some sort of motivational talk—because, after all, what would the rep of a clothing manufacturer have to say about soil health?

A lot, as it turned out—and I also found myself quite motivated after listening to him. He was introduced as his company's director of sustainability, and he spoke about corporate efforts to encourage and incentivize cotton farmers to use cover crops, less tillage, more intensive crop rotation, and other practices that go far toward preventing depletion of the soil.

My jaw dropped. For two decades, I had been advocating cover crops to dramatically improve soil health. It was the right thing to do. And now I saw another perspective: it was also the profitable thing to do. I could advocate cover crops to dramatically improve business health as well as soil health. What's good for the earth is also good for the market, and it was no longer just a niche market.

This was a breakthrough moment for me. I call it my epiphany. As I listened, it became clear to me that the major commodity producers—the growers and buyers of cotton, corn, soybeans—were feeling the influence of an increasing public demand and responding to it. *This is huge*, I thought. *Cover crops have gone mainstream.*

Atwood invited me to visit Wrangler's headquarters in Greensboro, North Carolina, and give a presentation, which I did the next

April. It was the same day that Wrangler was hosting a daylong soil-health conference, primarily for the Future Farmers of America, with which the company has been associated for more than half a century. These young people were there to learn specifically about the value of cover crops, rotation, and no-till planting. We also visited cover crop test plots at North Carolina A&T State University.

I asked Atwood why the company was taking an even closer look at the value of such practices. "Why are you going down that road?" I asked.

"We do it for the American farmer," he said. "We're sending a signal to our cotton growers that being a good steward of the land makes financial sense, too." Wrangler has invested in a series of soil-health trainings and farm trials to demonstrate that cover crops, rotation, and alternatives to tillage are both feasible and lucrative for farmers, with long-term benefits. Originally, he said, the innovative way that ranchers practiced rotational grazing inspired the company to

Wrangler invited me to participate in some soil health educational events.

increase its dedication to soil health. Wrangler has concluded that regenerative farming is both the profitable thing to do and the right thing to do. What is good for the market is also good for the earth.

That is the reason, then, that I have since been focusing on the power of the marketplace as I promote cover crops. The promise of potential profit will always be a strong incentive to take action. To this day the most common question that farmers ask, in survey after survey that I have seen assessing their interest in cover crops, is this: Do cover crops pay? It's a question that I, too, asked back in the mid-'90s before conducting research on my own farm. I was convinced, as I'm sure all farmers can be convinced.

CORPORATE INITIATIVES

No doubt about it. The big players are paying attention. They see what the market will increasingly demand. They recognize that unless they get with the game, they will miss out on the opportunity for some big profits. That's not to say they care only about money—they are as sincerely sensitive to environmental issues as anyone. Clearly, though, they perceive that consumers will reward their efforts to do more.

> No doubt about it. The big players are paying attention.

Never mind the cynical few who call those efforts "greenwashing" and dismiss them as more about marketing than about caring. The big agricultural companies have a lot at stake. They are aware that climate patterns, for whatever reasons, have shifted, and they see a potential for disruption of their supply chains and production. They have every reason to position themselves to protect the land that nourishes both the crops and their business.

Toward that end, major corporations whose fortunes are tied to the land have been focusing on a range of progressive practices. The soil, enriched with organic matter from cover crops and undisturbed by tillage, holds more water and releases no carbon into the atmosphere. In fact, cover crops take carbon dioxide out of the atmosphere and deposit it into the soil as carbon, a clear win for the farmer and the climate. The control of greenhouse gas emissions has been a key corporate initiative, as well as reduced food waste, more composting, and managed grazing of farmlands. Companies also have catered to the increasing popular demand for organic and non-GMO product lines. They have invested in global efforts to rebuild biodiversity and to halt deforestation.

Such efforts are widespread and rooted in both corporate self-interest and a concern for the public interest. Every chapter of this book returns to that theme of the growing alignment of business and environmental concerns. They work together, in symbiosis, much like a cover crop works so well with a cash crop.

I could cite so many examples of corporate contributions to good farming that this book would never end—so let me give you a sampling of a few that have been in the news, or on my mind, as I write this:

Bonduelle

In Chapter 1, I mentioned that I had consulted with Bonduelle on cover crops and no-till. The interest that this huge international corporation is showing in sustainable farming should decisively answer the question of whether it's just some passing fad or the way of the future.

Bonduelle has forty-two vegetable processing factories around the world, including several in the United States and Canada. I visited one of the company's factories in Hungary, where I talked to nine field managers, who work directly with the farmers, about cover crops and no-till. Then I traveled to a 4,400-acre farm in Hungary that grows sweet corn, peas, and dry beans for the company. A hundred acres of that land will soon be devoted to regenerative methods—not bad, for a start, as the farmers figure out how to transition.

Training Bonduelle field managers in Hungary.

As I entered the conference room of that farm, I noticed two things right away: six inches of padding on the doors and three large plaques on the walls. One of the farm business partners explained to me that the plaques had been awarded by former Soviet leader Mikhail Gorbachev acknowledging that the farm met official requirements. "And those padded doors?" I asked. He told me they, too, dated to the Soviet era, when the room was soundproofed for secret agricultural meetings.

My thoughts turned to President Reagan's challenge to Gorbachev to "tear down this wall!" As the Berlin Wall crumbled in 1989, so did the Iron Curtain in Eastern Europe. I asked him how he felt about Reagan's actions. His response was quick: "We are very happy Reagan did that." That brief exchange brought back boyhood memories of the Cold War and the Soviet threat from the other side of the globe. I felt humbled now to be helping the good people of this former Communist Bloc nation to grow better food.

Cover crops and no-till farming clearly have the support of Bonduelle's top brass. While I was in Hungary, Jean-Marie Sol, a representative from the company's headquarters in France, joined me to support my message. The company's leadership is clearly pushing for a global initiative to grow its food in a more environmentally friendly way. A company doesn't get that large without paying intent attention to marketing. Bonduelle has felt the shift in the wind, and it wishes to do its part for the planet. In short, this world leader in food production has concluded that cover crops pay.

With the father and son co-owners of a 4,400 acre vegetable farm in Hungary that grows for Bonduelle. Notice the 1980s awards on the wall from the former president of the Soviet Union Mikhail Gorbachev.

General Mills

The packaged food giant announced in March 2019 that it would advance the practice of regenerative agriculture on one million acres of farmland by 2030, or about a quarter of the land from which it sources ingredients in North America.

Citing estimates that a third of the planet's greenhouse emissions come from the food system, General Mills said its efforts would help to pull carbon from the air and store it in the soil. The goal, it said, was to fight climate change and improve food and water quality as well as help farmers to increase profits.

Through regenerative practices, the company pointed out, farmers disturb the soil less, keep it covered, maximize crop diversity, integrate livestock into crop production, and maintain living roots year-round to enrich the soil. The company will partner with key suppliers for such ingredients as oats, wheat, corn, dairy feed, and sugar beets. It intends to sponsor on-farm training and education academies where growers can learn the techniques. Since 2015, General Mills has invested more than $4 million to advance soil-health initiatives. It previously announced a commitment to reduce its greenhouse gas footprint by 28 percent by 2025. The company said it was about halfway to that goal, with its emissions in 2018 down by 13 percent from its levels in 2010 for all its business-related activities.[5]

---

5   "General Mills to advance regenerative agriculture practices on one million acres of farmland by 2030," General Mills news release, March 4, 2019, https://www.generalmills.com/en/News/NewsReleases/Library/2019/March/Regen-Ag.

"One Planet" Biodiversity Coalition

Nineteen leading companies—including Danone, Kellogg, Nestlé, Mars Inc., McCain Foods, and Google—recently joined forces to launch an initiative to advance regenerative agriculture, rebuild biodiversity, and halt deforestation.

The announcement of the "One Planet Business for Biodiversity" coalition came in September 2019 during the United Nations Climate Action Summit in New York City. The nineteen companies sell products in more than 120 countries and have combined total revenues of about $500 billion.

The coalition members will be focusing on devoting resources to three main areas.

- They will increase the use of regenerative agricultural practices. The goals include keeping carbon in the soil, increasing the soil's capacity to hold water, growing crops richer in nutrients, using fewer synthetic fertilizers and pesticides, and supporting farmers' livelihoods.

- They will develop products that expand the biodiversity of fields under cultivation, with more emphasis on local sourcing. A wider variety of ingredients will mean less reliance on just a few crops.

- They will work to eliminate deforestation and to restore and protect ecosystems of high value, including grasslands, wetlands, and forests.[6]

---

6   "Nineteen leading companies join forces to step up alternative farming practices and protect biodiversity, for the benefit of planet and people," World Business Council for Sustainable Development news release, September 23, 2019, https://www.wbcsd.org/Programs/Food-Land-Water/News/Nineteen-leading-companies-join-forces-to-step-up-alternative-farming-practices-and-protect-biodiversity-for-the-benefit-of-planet-and-people.

Shell Oil

The regenerative agriculture movement has even caught the attention of Shell Oil. The petroleum giant funded a study on intensive grazing of grasslands in Alberta, Canada. More and more research worldwide has been looking for ways to enrich soil that has been depleted of carbon. The aim is to help rangelands gain the resiliency to withstand increasingly severe droughts.

The Alberta study examined the soil on ranch paddocks where cattle grazed continuously and compared it with the soil of paddocks where the grazing was heavy for short periods between long recoveries. The latter condition simulates natural ecosystems that resulted from the migration of bison, caribou, and other species. The study found that intensive grazing increased the organic carbon sequestered in the soil by a statistically significant amount and also increased the rate of water infiltration.

The study was funded by Shell GameChanger through a grant to Arizona State University. *GameChanger* provides seed money and support for initiatives that explore ideas and technologies that could transform the energy industry.[7]

THE GREAT MOTIVATOR

Since about 2016, I have changed how I answer a perennial question. "What will it take to get more farmers to plant cover crops?" people often have asked me. Until then I generally mentioned three possibilities. One would be educating growers so that they see the benefits and act in their own best interest. Another would be imposing government regulations, which is hardly a popular option—a tough row

---

7 "Alberta Soil Carbon," Applied Ecological Services, accessed May 11, 2020, https://www.appliedeco.com/portfolio/alberta-soil-carbon/.

to hoe, so to speak. And the third would simply be a spike in fertilizer prices, for whatever reason, which would incentivize farmers to plant cover crops as a cheaper way to boost nitrogen levels. The demand for cover crop seeds would skyrocket.

Notice the nodules on the roots of a sunn hemp cover crop? Legume plants take nitrogen out of the air and store it on the roots, which lessens the need for farmers to purchase it!

All of those would move the needle in favor of cover crops—but from my perspective, as someone working to advance the cause, I can control only the educational element. I can't wave a wand to raise fertilizer prices. I certainly can't issue orders telling farmers what they must plant, nor would I ever dream of doing that. They need to see for themselves what makes sense and to use their own best judgment based on what they have learned. Once they give cover crops a try,

I'm confident they will conclude that they should have been planting them all along.

Farmers also can see for themselves what I have concluded from the interest that Bonduelle and Wrangler and General Mills and all the others have shown. The evidence is clear that corporate eyes are on the changing market—and today, that's how I answer the question about what is most likely to move the needle for cover crops. The incentive from the market will persuade increasing numbers of farmers not only to cultivate cover crops but also to learn to do it right so they can keep their competitive edge. Making more money is a great motivator for anyone, whether it's the big guys or the small farmer just trying to get by.

For those on the fence, the big question is not so much whether cover crops pay but rather when to get started. Time is money. Delays are costly. Farmers know that. Anyone with experience in business knows that. I do understand why farmers have their reservations about whether it's worthwhile to make the switch. I did, too. Now I don't.

Back in '99, it was an increased yield of corn that turned me into a cover crop devotee. That was after four years of trying cover crops on several test plots on my farm—research that I had been conducting with Dr. Weil, the world-renowned soil scientist from the University of Maryland. Until then, I had been publicly posing the question that perhaps no-till practices alone might keep the soil in sufficient shape over the long term. I wondered whether the chief value of cover crops was just to help the transition to no-till.

After those four years, however, I wasn't wondering. I had my answer. The research did the talking. That year was dry, and those test plots were producing twenty-eight more bushels per acre of corn than the control plots where we didn't grow cover crops. And the earthworms in the test plots were wriggling in delight.

*Dr. Ray Weil on my farm evaluating good soil
health achieved by using cover crops, no-tillage,
and plant diversity. Notice the earthworm?*

It's high time to get on board. We know that cover crops can transform the business of farming. The benefits have been demonstrated time and again. The next question is this: How do we make the most of those benefits?

It's a matter of commitment and good management. A farmer might agree that cover crops are the way to go, but still needs to earn a living. That is why I so often hear the refrain of "Do cover crops pay?" The farmer wants to do the right thing, but his immediate concern is providing for his family. What I offer in this book is reassurance, based on the preponderance of evidence, that he can do both. There are compelling reasons for committing—and it goes

> There are compelling reasons for committing—and it goes without saying that sound management must be part of the game plan.

without saying that sound management must be part of the game plan.

POSITIONING FOR SUCCESS

Effective management of any sort requires time, knowledge, and dedication. Farmers understand that implicitly. As they plan their cash crops, they do their due diligence. They shop around to keep costs down for seed and fertilizer. They look for the best deals on equipment. They buy what they can afford. This is what farmers do year in and year out, season by season, as they manage their cash crops. They are comfortable with making such decisions. They do their research, make the projections, count the costs, and do what is best for their business. That's how they manage their cash crops. They do it so that they have money in the bank at year's end.

Cover crops are a different beast that requires a different tack on management. "I know they're a good thing," farmers tell me, "but what about the up-front cost of the seeds, and how can I justify the time and expense of those extra trips across the field to get the crops established?"

I can suggest ways that they can start out with more bang for their buck. For example, one tactic is to grow a small grain in the rotation and then plant a mixed-species cover crop for the summer. The cover crop will significantly reduce the farmer's fertilizer bill the following year, thereby making up for the cost of the seed. Some farmers are in a position to grow their own cover crop seed, such as cereal rye or triticale, making the transition more cost-effective on a low budget. Or they might be able to seed their cover crops at a lower rate than recommended and still get great growth through careful management and timing. For farmers with livestock such as cattle or sheep, another tactic is to graze the animals on the cover crop. That's pretty much a guaranteed return on investment right away.

It's true, though, that you might not see a quick, immediate return from cover crops—but that's the nature of many investments. An impressive gain at the start is more the exception than the rule. Mostly, your investments grow over time as you plan for a bright future. Each year's growth builds on the previous year's discipline. If you are planning to add cover crops to your rotation, find an experienced mentor who can help you maintain that discipline. A mentor can help you avoid costly mistakes, learn what works and what doesn't, and see the profits sooner.

In other words, position yourself for success, as those big corporations are doing. The individual farmer likewise must plan long term. Cover crops are not a one-year proposition, in the way that farmers tend to think of their cash crops. A farmer can calculate at the end of the year, based on earnings and expenses, whether a planting of corn, for example, has been profitable. By contrast, I tell farmers to think of cover crops as a ten-year proposition. The better they manage their investment in year one, the greater the payoff by year five. After a decade, and probably sooner, they will be wondering why they didn't start much earlier to cash in on a priceless opportunity.

I have done my due diligence, and all the indicators say full speed ahead. After years of experience on my own farm with cover crops, I haven't the slightest doubt about their value. For anyone in the know, the question of whether they pay is pretty much moot at this point. Yes, they do, and I'm not talking about getting subsidies for trying them out for a few years. I'm talking about their inherent value over the long term for those who dedicate themselves to learning all they can.

I'm sure that the veteran cover croppers out there who can look back at the difference in their fields over five or ten years will agree wholeheartedly with me. Cover crops enrich the soil and fatten the wallet. They are good for the environment and good for business. They simply are an essential of good farming—and good farming pays big dividends.

CHAPTER 5
THE GOOD EARTH

THE GOOD EARTH

My dear wife, Cheri, grew up in Costa Rica, the daughter of missionaries, and in so many ways she has enriched my life, including with the cilantro and mangoes and other delights that she introduced to our table. But this Pennsylvania farm boy most enjoys her simple touch with the squash we grow here at Cedar Meadow. When she serves it hot from the oven with butter and a sprinkling of brown sugar … *wow!*

We didn't bother with the condiments, though, after tasting a new variety of squash called Koginut that we began growing back in 2018 for the Sweetgreen salad restaurants. It's like butternut but even better. The flavor is so rich and smooth that it's best without anything getting in the way. I tell Cheri she doesn't need makeup to be perfect to me. She doesn't need help. It's sort of the same idea with Koginut. When you've got something good, don't mess with it.

I met Cheri in 1986, a few years after graduating from high school. She was living in Texas, where her parents had relocated after their long service in Costa Rica, and we were both attending a six-week Bible school in Ohio. At our first meal there, Cheri and I took a seat at the same table (by divine appointment, I'm sure) and struck up a conversation. As the weeks went by, we began hanging out together more and more—until the school concluded, and suddenly half a continent separated us again. We spent a lot of time talking on the phone and writing letters, which some of us remember is what folks did before these days of texting, emails, and Instagram. Eventually, she paid a visit to our farm in Lancaster County. Her mom and dad had grown up on a Kansas wheat farm, but that kind of lifestyle was pretty much new territory for a girl who spent the first sixteen years of her life in the tropics.

Cheri's Koginut squash—ready to eat!

Two years after we met, I asked Cheri to marry me. We had been getting together every couple of months, and during one of my trips to San Antonio, at Christmastime, I felt that the time might be right. I wasn't sure, but I'd been rehearsing my line. Her parents had a backyard treehouse, about ten feet off the ground, and as midnight approached on Christmas Eve we climbed up into it. When we heard fireworks in the distance, I knew that the hour had struck and took her hand and posed the question. "Yes, Steve, I will!" she said, and we sealed the deal with a kiss as the fireworks kept popping. In keeping with my conservative Mennonite upbringing, we skipped the engagement ring. Instead I bought her a piano, which she still plays regularly in our living room. An anniversary ring came later.

Within six months we married and set up household in the 1890s farmhouse where my grandparents had lived—what we call the "home farm." And though Cheri tells me that she fell in love with the farmer, not the farm, she grew to appreciate the life that we would forge together, raising our family here on the good earth of southeast Pennsylvania.

What most people picture when they think of Lancaster County isn't what we have in our neck of the woods, and I mean that literally: forests cover half the land here, about fifteen miles south of Lancaster, Intercourse, Strasburg, Bird-in-Hand, and the other well-traveled tourist towns to our north. We have more hills than they do—and one of those hills is a field that comes to within a hundred feet of our farmhouse.

On our land we grow the variety of crops that earn us our living, along with a herd of a dozen buffalo that we graze on a six-acre meadow just below the barn. And among those crops is the fabulously flavorful Koginut squash.

Koginut squash developed by Row 7 Seeds and grown for Sweetgreen.

FROM SEED TO SALAD

During a holiday trip with my family, in that week between Christmas and New Year's when life seems suspended, we stopped at a Sweetgreen location in Alexandria, Virginia, just south of Washington, D.C. It was busy that day, and as we waited our turn, I noticed a chalkboard that listed the thirty or so farms that supplied produce to the restaurant chain. At the top of the list: "Cedar Meadow Farm, Pennsylvania, Koginut squash."

I looked through a customer brochure titled "The Future Is Flavor," and it highlighted our farm as one of only six in the United States that grew that variety. I felt proud of our contribution. "Let me tell you who I am," I told our server as we watched her prepare our order. "I'm the farmer who grew your Koginut squash."

"Wow! You're like a hero to us." Her expression was one of genuine delight.

Me pointing to Cedar Meadow Farm
listed on the board at Sweetgreen.

After we finished our salads, I paused for a photograph of the chalkboard sign, but I couldn't get a good shot because of the crowded serving line. I tapped the shoulder of a tall gentleman and asked him to help me out by holding back the line for a moment. I explained why I wanted a picture of me standing in front of the sign.

"Sure," he said, studying me a moment. "You're the farmer?" he asked, smiling broadly as he reached out to shake my hand. "I've never met a farmer before. Thanks for growing our food!"

His appreciation was genuine, and I was hearing it directly from someone who was buying and enjoying what I had grown. *That would sure never happen at the grain elevator*, I thought.

We began growing the Koginut squash at Cedar Meadow in 2018 at the request of Sweetgreen, which wanted to include this new variety in its menu. Founded in 2007, Sweetgreen has dozens of locations across the nation and calls itself "a destination for simple, seasonal, healthy food." Committed to transparency, the company strives to ensure that its customers know where their food came from and that they see how it is prepared.

The Koginut squash is a creation of the Row 7 Seed Company, founded by world-renowned chef Dan Barber, vegetable breeder Michael Mazourek, and seedsman Matthew Goldfarb. They wanted to form a company "dedicated to deliciousness." Row 7 describes itself as the first seed company in which chef and breeder collaborate to make an impact on the soil and at the table. The company has grown to become a nationwide distributor of specialty seeds.

"One little squash started it all," the Row 7 website explains. Barber joined forces with Mazourek, a Cornell University professor, to develop the honeynut squash. In search of a hybrid that would combine the richness of butternut squash with the smooth taste of Japanese kabocha, Mazourek designed the incomparable Koginut.

Mazourek was pleased to be breeding for flavor. Growers typically want plant breeders to select for yield or shelf life or uniformity, at the expense of taste and nutrition and the environment. Row 7 chose instead to put the focus on delicious. The company points out that the compounds that enhance flavor and aroma often derive from essential nutrients that are present in the produce. That's nature's way of rewarding us for eating right.

At Barber's highly acclaimed restaurant, Blue Hill at Stone Barns, near Tarrytown in New York, some of the cuisine has been grown from Row 7 seeds. The restaurant uses all local ingredients from farms in the Hudson Valley, including produce grown in the extensive gardens surrounding the refurbished barn, once part of the Rockefeller estate. Blue Hill at Stone Barns is one of a select few restaurants ranked at three stars in the *Michelin Guide*. Since 2015, it has consistently made the list of the World's 50 Best Restaurants, compiled by a panel of international chefs, restaurateurs, and food critics.

That's quite a reputation to uphold, and Barber does it simply and delectably. I know because I experienced it. At the chef's personal invitation, Cheri and I had the honor of touring Blue Hill and dining there. It was like visiting the Grand Canyon: you know it's going to be good, but the experience is even grander than you ever imagined. To get to the restaurant, we drove through the fields that produced what we would soon be enjoying. Over the course of four hours, we were treated to a couple dozen artfully presented servings, each of which celebrated the bounty of the land.

Among the fare were varieties of cover crops, including Tillage Radish, a cover crop that I originally developed. We enjoyed winter peas, crimson clover, hairy vetch, little sprigs of buckwheat, and much more. For dessert, we were ushered to a table in the bustling kitchen, where Barber took the time to talk with us during the thick of a busy evening. We shared our own story and dreams and found out more about his.

Cheri and I with Chef Dan Barber in the kitchen of his restaurant, Blue Hill at Stone Barns.

Barber and I have similar goals of growing highly nutritious food and spreading awareness about the use of cover crops. That's why he has been including them on his menu and explaining their value. In the weeks after our exquisite dining experience, Barber's growers contacted me with questions on implementing cover crop strategies. That was a first for me—to be consulted about growing cover crops destined for human consumption.

Meanwhile, Sweetgreen also has been developing a similar philosophy of nutritious food, from farm to table. I got a call out of the blue from Sweetgreen one morning in September 2017: "We heard about your farm and how you grow with cover crops and no-till. Would you mind if we stop by to talk about maybe growing some produce for our restaurants?" I'm always open to new markets, so

I agreed to meet with the company's representative, Allison, who offered to come out to our farm.

"We're looking for farmers to grow a specialty squash called Koginut," she told me. This was the variety that Mazourek, the Cornell vegetable breeder, had developed for Row 7, working with Barber. Sweetgreen wanted to add Koginut to its menu because it met the criteria of delicious and nutritious. The company put me in touch with Mazourek, and I sent him an email asking for his expertise on how best to grow this variety, such as the recommended plant spacing and fertilizer requirements. "I know who you are," he responded. "I heard your talk when you were up here for the New York vegetable growers' convention."

I was impressed by Sweetgreen's overture. This was the first company that had ever approached me to ask if I would grow for its customers exactly what I was most interested in growing myself—that is, highly nutritious produce that would pique the public's appetite for more farm-to-table fare. My farm was one of only six across the country that Sweetgreen selected to grow Koginut squash for all its locations in the East. In that first test year, the company even footed the price of the seed. It accepted some of the farmers' risk in trying out a crop so new to the scene. As it turned out, the Koginut flourished at Cedar Meadow Farm.

The Koginut venture has been a win for all. Sweetgreen's interests were aligned with my interests, which were aligned with the interests of the chef, the vegetable breeder, the seed company, the consumers, and anyone interested in good food and good farming. As the folks at Sweetgreen put it: "We're building a community of people who support real food."

"O TASTE AND SEE ... "

Sweetgreen had heard about me because of my long history with cover crops and no-till farming.

In the early 2000s, Ed Huling, a USDA researcher in Beltsville, Maryland, tested some of the produce from my no-till and cover-cropped fields. The results clearly showed my veggies had a higher nutritional density than the USDA average—meaning more nutrients in every bite. I have kept in contact with him over the years, working on some projects that could lead to reducing or eliminating the use of fungicides on my crops. One of the perks that we and others have seen with nutrient-dense fruits and vegetables is that they can better resist disease and insects.

I also was influenced by the work of John Kempf, who went on from his eight grades of Amish schooling to become one of the sharpest minds in agriculture. He teaches how a comprehensive systems-based approach to plant nutrition will pack usable nutrients into the food we eat, resulting in better human health. The theory is simple, but getting tangible results is a complex task—and a challenge that he has eagerly embraced.

I wasn't alone in those initiatives, having met some innovative farmers over the years. In 2003, for example, Washington State farmer Karl Kupers cofounded Shepherd's Grain, which today represents about forty growers who farm sustainably. Kupers wanted to establish through testing that his no-till farming enriched the soil to produce more nutritious wheat. The local bakers were already attesting to the fact that the flour from his wheat produced tastier bread, so he decided to put some science behind that observation. He confirmed the higher nutrition levels with the help of Canadian soil-health researcher Jill Clapperton, an international lecturer who has twice attended field days at Cedar Meadow Farm to share her work.

Despite what the facts were showing, however, I soon discovered that few others back then seemed all that interested. The stores weren't paying any kind of a premium. Still, I knew that quantity couldn't be everything. Surely consumers would be demonstrating with their purchasing power that they cared about quality and nutrition—and that prediction has largely come true.

> # Surely consumers would be demonstrating with their purchasing power that they cared about quality and nutrition—and that prediction has largely come true.

Long before the farm-to-table movement was in vogue, I prioritized balancing the soil nutrients in my fields. Farmers tend to focus on the big three, nitrogen, phosphorus, and potassium, which go far toward increasing yield. Calcium, sulfur, boron, and zinc also play important roles in boosting yield. Farmers tend to pay less attention to a variety of micronutrients that advance quality but don't do much for yield. Why bother when you're paid by the bushel?

I bothered. I began applying more broad-spectrum fertilizers, including naturally mined products such as aragonite, azomite, and a few others that I believed were boosting the quality of my squash. I conducted some tests, and sure enough my squash registered higher than the United States Department of Agriculture average for virtually every nutrient.

I concluded, though, that the micronutrients were only part of the equation. My use of no-till and cover crops through the years had much to do with generating a better, tastier, healthier crop. Those methods play a big part in unlocking nutrients that occur naturally in the soil. Specialty fertilizers added to the soil certainly help, but much of the improvement comes from observing sensible farming practices that have produced good food for countless generations. I realized back then that if more farmers changed their ways, they could release the nutrients in the soil that were there all along. In other words, farmers could wake up the soil—if they would just wake up to the opportunity.

In 2017, the year before I began growing Koginut, I had the lab run another test on my butternut squash. Once again, it was significantly higher in every nutrient except sodium, and to be lower in sodium isn't such a bad thing. All in all, those two tests, about fifteen years apart, were telling me the same thing: the soil on my farm was delivering superior results from the same varieties of seeds that had been planted on other farms. I had bested the nutrition levels normally expected of those seeds. Regardless of whether my farming practices or the specialty fertilizers were the reason, I was pretty pleased.

The big change over those fifteen years was that the buying public's demand for quality had become undeniably evident. Nonetheless, farmers still, by and large, are paid on the basis of quantity. Their profits come from higher yields, so adding certain supplements can seem like an unnecessary strain on an already slim margin. Why add copper or silicon or selenium to the soil, for example, if it doesn't raise the yield or market price?

To that question, I make three observations:

- As demand rises, so do prices. We have already established that demand is rising for more nutritious food.

- Farmers who plant cover crops without tillage can consistently maintain or even improve yields even if they use less fertilizer. Meanwhile, they are conditioning the soil to release its natural nutrients and hedging against weather extremes.

- Nutrients are good for people's health. There are many reasons for chronic health problems in this country, and one is poor diet. Farmers can help by providing nutrient-dense food.

The attitude of indifference toward food quality has been changing dramatically, as evidenced by the very existence of Sweetgreen, the Row 7 Seed Company, and the numerous other ventures that have taken the pulse of the public in recent years. Individual farmers who see what's coming have been joining that trend toward better nutrition. It's a market opportunity not to be missed.

To meet the rising demand, farmers must take responsibility for the soil. When the soil is enriched, so is whatever they grow in it. However, consumers who care about quality are looking for more than good nutrition. They want exceptional taste, too, and that's something they can't get from a pill.

For centuries, that taste has been attributed to the quality of the soil. The concept of *gôut de terroir* (which means "the taste of the earth" in French) has been applied to a variety of artisanal crops that are deemed to have a unique regional character. Today, people prize such products as coffee, tomatoes, cheese, tobacco, and cannabis for their regional qualities.

According to legend, the monks of Burgundy in the Middle Ages licked the dirt to judge whether it was worthy of winemaking.[8] I'm guessing they were probably brighter than that, but the tale does show that folks have long perceived a relationship between soil and the flavor of what is grown in it. As for me, I don't lick the soil, but I sometimes smell it to judge its health.

---

8   Mark A. Matthews, *Terroir and Other Myths of Winegrowing* (Berkeley: University of California Press, 2016), 179-180.

Smelling the soil in Uruguay.
A healthy soil will cause you to take a second sniff!

In my travels, I have found that the emphasis on local flavor is much more pronounced in Europe than in America. At a farm meeting in France, for example, I might be served a specialty cheese along with a glowing account of its pedigree and the region where it was produced. At a farm meeting in America, the caterer more likely will serve some generic cheese—delicious, perhaps, but without fanfare. Europeans so often have a rich story that goes with their food. They show that they care about where and how it is produced. In America, a lot of people seem to regard food with a shrug.

That attitude, however, has clearly been changing for the better in many circles. The staff at Blue Hill at Stone Barns, for example, took the time to explain each serving and to answer questions.

Cheri and I found them to be highly knowledgeable about farming, sourcing, and food preparation. We got a vivid picture of the *who, what, when, and where,* and we left without any doubt about the *why.*

In other words, there are stories to tell, and at Stone Barns they tell them well. Along with simple excellence, the emphasis there is on education. The Stone Barns Center for Food and Agriculture is an integral part of a mission "to create a consciousness about the effect of everyday food choices." Stories help to promote that awareness. They clarify concepts. They engage people and get them excited about possibilities. They motivate.

Cherry tomatoes ready to ship to Sweetgreen.

The Sweetgreen folks likewise have dedicated themselves to raising awareness. Their mission, too, is farm-to-table transparency, and they have been taking it to a new level. After adding tomatoes to their wish list from me, they came out to Cedar Meadow Farm to put probes and sensors into the soil where I plant those tomatoes. The probes monitor a range of data throughout the year including weather and moisture conditions. In a nationwide taste test, Sweetgreen found that its customers rated my tomatoes highest. There is tangible evidence that nutritional quality improves taste, and I look forward to the day when the science proves it.

I also agreed to supply Sweetgreen with my planting, harvest, packing, and shipping dates. Presumably, customers will soon be able to hold their phones up to a QR code or use some similar method to get a readout of where the tomato and other produce in their order was grown, when it was planted and harvested, the growing conditions, and a lot more. I still kind of like Sweetgreen's homey chalkboard, but the high-tech blockchain system does indeed offer a clearer picture of what goes on all the way from seed to salad. It tells a better story.

That's all in keeping with the Sweetgreen ethos of openness and connection between farmer and consumer. The focus is shifting to letting the consumer be aware, rather than just beware. Information is power. I think again of the nutrient-measuring device that I recently tried out in beta testing that conceivably could let supermarket shoppers one day measure nutrients with a tap on an app. Though it may be too early to tell whether the consumer monitoring model will become the standard, I certainly am intrigued. And I want to partner with people who think that way. I want to partner with people who look to the future with respect for the past. My aspiration is to come alongside them and offer whatever I can to help set them up for success.

This is a truth that I wish to share with my colleagues who care about the land and about our way of life: by telling our stories and spreading the good word, we will draw closer to those who value what we offer. They will see for themselves the bounty of the soil, and they will taste what we have planted, grown, harvested, and prepared with simple excellence. I think of these beautiful words from the Psalms: "O taste and see that the Lord is good."

A HEALTHY PERSPECTIVE

I have seen multivitamin promotions that suggest you should pop a daily pill to replace what farmers don't provide anymore. The implication is that agriculture has depleted the soil to the point where we can no longer maintain healthy bodies naturally and need to take those supplements.

You could choose to do that, of course. You could pop a pill to get the recommended daily doses of the nutrients you need. Likewise, farmers could dump fertilizer into the soil to give it the nutrients it should have. That covers up the problem for a while, but in the long run our society needs something inherently healthier, something more foundational.

I'm not out to put the multivitamin companies out of business, but I will say there's a better way. By eating a variety of good, nutritious food, you can get what you need in your diet without buying those expensive supplements. Everyone seems to agree with the multivitamin makers that our food is not as nutritious as it once was. The solution shouldn't come in a bottle, though. We need to directly address that deficiency in our food.

It all starts with regenerating the soil. It will produce more nutritious food when we allow it to do its job. To enrich the soil, farmers must disturb it as little as possible while introducing a greater diversity of plant life. Throughout the year, living roots should be pushing through the soil. By giving it a chance to function again as nature intended, we can restore its power.

The solution isn't all that mysterious. It's simply a matter of respecting nature. God gave us a diversity of plant life to cover and protect the soil with green growth. If all humans were to vanish, the plant life would still thrive. Through the cycle of the four seasons, the roots would continue to penetrate deep into the soil to replenish it and bring forth a profusion of life. The greenery doesn't need us. We, however, need the greenery.

As farmers try to feed a hungry world, they are at their best when they mimic nature. They will produce the best crops when they treat the soil as more than a depository for chemicals. Long before the advent of agriculture, the soil was already doing a great job of supporting growth. Nature hasn't lost that ability. It can still generate nutrients aplenty—so long as it is alive. Doctors can keep people going past their expiration date by pumping stuff into them through tubes, but that's not truly living. Nor should our soil be on life support. It is the farmers' duty to keep the soil healthy, functioning naturally, and truly living.

Volumes have been written on the technicalities of soil management, but all those details come down to a simple principle: better soil leads to better food, which promotes better health. Much of the research into those specific connections is still in its early days, with so many variables besides diet to consider, such as genetics and environment.

> All those details come down to a simple principle: better soil leads to better food, which promotes better health.

The anecdotal evidence seems clear, though. Some farmers have observed that their cattle or their hogs are healthier, or go to market sooner, when fed on corn from no-till fields where cover crops were grown and fewer fertilizers and pesticides were applied. An Amish neighbor told me his cows "milked better" when he fed them the corn silage that I grew for him. And in central Ohio, David Brandt farms over a thousand acres with no-till and cover crops. We've been to each other's farms several times, sharing notes on how to help other farmers convert to regenerative agriculture. Brandt sold corn to a neighbor who reported that his hogs reached market a few days sooner and needed less medication. The neighbor was impressed enough to offer him a few cents more a bushel for corn grown that way.

Those farmers aren't claiming to be conducting scientific research. They just know what's going on in their fields and barns. The livestock are doing well. Their conclusion, plain and simple: *Must be something they're eating.*

If regenerative agriculture can help animals to thrive, it stands to reason that the human body, too, would benefit. The connection between soil health and human health has lately been the focus of national conferences, attracting hundreds of scientists and organization leaders. The topic is clearly timely. If we don't take care of the soil, we cannot hope to take care of ourselves.

Consider this: many of the microbes in the human digestive system can also be found in the soil. The microbes that we harbor in the gut provide a range of proven health benefits, and we need a diversity of them. Studies have found that growing up on a farm, where microbes are abundant, helps children to develop stronger immune systems. In the general population, however, the diversity of intestinal microbes has declined dramatically, the research has found. Among the likely reasons are a prevailing diet of processed foods and a more antiseptic urban lifestyle. This microbial imbalance could have major implications for public health. [9]

If we agree that it matters what we put into our stomachs, then we also can agree that it matters what we put into the soil. Soil health

---

9   Winfried E.H. Blum, Sophie Zechmeister-Boltenstern, and Katharina M. Keiblinger, "Does Soil Contribute to the Human Gut Microbiome?" *Microorganisms* 7(9), August 23, 2019, 287.

is a matter of good management. It doesn't come in a bag. It's not something you buy. Soil management is something you do, day by day, until you get it right. You feel the rhythm and learn the steps. If you do a good job as caretaker of your land, the soil will be richer. It will have greater biodiversity, more organic matter, more earthworms and microbes, more nutrients. It will function better. It will grow better food for a healthier society.

I have done plenty of research in my own fields, but I must point out that it only goes so far. Research results can be confusing. All those variables can be maddening. It's difficult to determine what causes what and why, and out of context, data can lie. Folks tend to pick and choose whichever truths seem to support their cause. Marketers have a way of zeroing in on whichever facts serve their own purposes while ignoring those that don't.

And it's a dirty little secret that the people who fund a research study often get the results they want. I've seen it happen. Whenever someone presents data and makes a claim, my first question is: "Who paid for that?" I'm not saying the findings in such cases are nonsense, necessarily. They usually include a kernel of truth. Always consider the value of the other guy's ideas—but maintain a healthy skepticism.

Agriculture has always been more art than science, anyway. I do admire the brilliant minds who have done wonders in advancing our knowledge and helping to improve our agricultural practices. It is no exaggeration to say that their contributions have fended off world starvation. They have greatly improved the public health. We owe a debt of gratitude to science, even though it sometimes is too slow to "prove" what farmers observe happening in their own fields. If we are to heal and regenerate the land, we need the practical sense of the practitioners of the art.

TECHNOLOGY AS A GAME CHANGER

Years ago, at one of the dozens of field days that I have held at Cedar Meadow Farm, I was taking a wagonload of visitors on a tour and stopped to give a talk about the benefits of no-till and to answer any questions.

"So do you use any herbicides yourself here?" one of the visitors asked.

"Yes, I do use herbicides at times to get rid of weeds and sometimes to terminate a cover crop," I said.

"Well, I like everything else you're doing here, Steve, but I have a problem with that."

I understood what he was telling me. If I was truly dedicated to no-till and cover crops as a natural means of regenerating the soil, why would I ever be willing to resort to chemicals? In response, I pointed to the hilly field where we had stopped.

"This field here has a 17 percent slope," I said. "So let me ask you this: how would you go about growing a crop on this field if you didn't use an herbicide? What would be the alternative for stopping the weeds?" I paused to let it sink in.

"Well, okay," he said. "I would have to till it."

"If you were to till the ground here, on this hillside," I asked, "would you expect to see quite a bit of erosion? And wouldn't the tillage kill earthworms and other critters that are beneficial to soil life?"

The gentleman nodded. "Yes," he said, "and I see your point." The no-till management of the crop on this field, which included the judicious application of an herbicide, prevented that erosion almost entirely. In this case, the benefit to the soil outweighed any issues with the chemical.

This was a matter of striking the right balance, like so much else in life. I fully agree with the goal of using fewer chemicals. Sometimes, though, as I explained to that visitor, they can be quite useful in pursuit of the overarching goal of returning the soil to its natural state of health. Ultimately, that new life is what reduces the need for chemicals, and that takes time.

Our society has long seemed to operate under the mentality that "if it's bad, we must kill it."

Farmers have tended to wake up each day thinking that way: *What must I kill today? A bug? A weed? A disease?* We think in terms of elimination more than proliferation. Here's a better question: *How can I bring more life to my farm?* As we go about finding that answer, we will discover, in time, that we have less to kill.

I'm not suggesting, though, that farmers should smile and let the armyworms and corn blight have their way with their crops as they wait for a better day. Just as I am not out to put the multivitamin makers out of business, I have no illusions that we will do away with synthetic fertilizers and pesticides. Farmers need them, but should think of them as supplements, not as replacements for nature's way.

The challenge is not how we can stop using the tools that science has given us, but rather how we can use them more wisely. Think of the automobile. It has done wonders to advance our society (though I fully respect my Amish neighbors who would beg to differ). Cars and trucks spew filth into the air, though, and crashes kill more than a million people around the world every year. They consume fossil fuels that we can never replace. Does that mean we should ban them for transportation and mimic nature by walking everywhere, maybe riding a horse? The pharmaceutical industry has given us the opioid epidemic that has killed hundreds of thousands of people. Does that mean we should regress to home remedies for everything that ails us? We shouldn't run to the doctor for every sniffle, but all and all, medicine serves us well.

One must always weigh the negatives against the positives, the risks against the gains. The automotive and pharmaceutical industries have transformed our way of life, but the cars and drugs that they produce are only tools. Tools can be redesigned and improved to serve both us and nature better. During the 1970s, when gasoline prices spiked and drivers waited in long lines to fill up, the automakers began developing more efficient vehicles to replace the gas guzzlers, at the same time reducing pollution. Today, research chemists are hard at work developing better, safer drugs. No one is suggesting we should shut down the pharmaceutical industry because of the opioid crisis.

> It is our responsibility to feed the world—working respectfully with nature, supplementing it as necessary, and constantly looking for ever better tools and methods.

Time after time, widespread attention on a troubling issue has led to a demand to find a better way. We don't give up. We innovate. And that is what the agricultural industry, like the others, must do. It's shortsighted to simply say we must stop using all pesticides and fertilizers because of the harm that they have done. We must consider the good that they have done, and then look for solutions. We gain nothing if we run away from the problem. Humanity cannot and will not go back to foraging for berries and nuts in the woods. Agriculture will rise to the challenge. It is our responsibility to feed the world—working respectfully with nature, supplementing it as necessary, and constantly looking for ever better tools and methods.

God gave us minds and a creative spirit. Just because something is man-made, or synthetic, does not mean it is bad or an affront to nature. Even when we determine that a practice has been harmful to the soil or the environment, we cannot immediately abandon it. Commercial agriculture today cannot survive on a philosophy of doing absolutely no harm. To produce food at a price that people can afford, farmers often must decide what will do the least harm as they await better innovations. The crops must be economically sustainable, too.

Any kind of growing method involves a degree of compromise, including organic agriculture. Organic farmers use pesticides too. They are derived from natural sources, but that doesn't necessarily make them safer than their synthetic counterparts. Some are known to kill pollinator insects. And most organic farmers still till the soil, which kills the life within it and subjects it to erosion. Organic farming generally is a good system, but it definitely is not the pinnacle of sustainability.

So yes, to this day I still use a modest amount of fertilizers and pesticides at my farm. I also buy gasoline and diesel fuel, and I take prescription drugs, albeit rarely and as a last resort. I strive for balance and moderation in everything. To those extremists out there who believe farmers are irresponsible if they continue to use any fertilizers and pesticides whatsoever, let me ask: Would you be willing to give up your car, electric power, or smartphone?

Technology has been a game changer. It has brought wonders to the modern world, and it will bring us solutions, too, as we identify whatever is the downside. One small example: my sprayer has a GPS control and field mapping software. It can sense my position in the field within an inch or two and keeps track of where I have been. If it determines that the soil has had just the right amount of whatever I am applying—fertilizer, pesticide, or microbes—it quickly shuts down the nozzles on the boom. No excess allowed. The technology saves me hours of time making those calculations and adjusting the machinery to prevent overlap. That's a benefit both to the environment and to my budget.

Only a few centuries ago, the simple cast-iron plow was the standard. It worked fine back East, but it kept clogging in the tougher turf and gummy topsoil of the Midwestern prairie. In the 1830s, John Deere invented the self-scouring steel moldboard plow.[10] His innovation dramatically increased food production—until, a century later, the prairie began to blow away as America's breadbasket turned into a dust bowl. In recent years we have come to understand the ravages of tillage. No-till and low-till methods, along with cover crops, are today's solution. They are helping to restore soil health, but we're not there yet.

Progress is an immense journey. We advance from one milestone to the next in search of better methods, better tools, better technology. Agriculture has come far, but the road is long. As we travel it together, learning and growing, we can face each new challenge as it comes.

STEWARDS OF THE LAND

Jesus of Nazareth used the power of parable to convey spiritual truths. "A farmer went out to sow his seed," he told the crowd gathered around him. "As he was scattering the seed, some fell along the path, and the birds came and ate it up. Some fell on rocky places, where

---

10   Marti Attoun, "John Deere: Agricultural Innovator," American Profile, April 17, 2005, https://americanprofile.com/articles/john-deere-pioneering-plow-maker/.

it did not have much soil. It sprang up quickly, because the soil was shallow. But when the sun came up, the plants were scorched, and they withered because they had no root. Other seed fell among thorns, which grew up and choked the plants.

"Still other seed fell on good soil," Jesus added, "where it produced a crop—a hundred, sixty or thirty times what was sown." He looked out on the crowd. "Whoever has ears," he said, "let them hear."

I could never hope to say it better than that. That simple story demonstrates that the success of our harvest depends on the quality of the soil—and of the heart. We wither from lack of roots and so many weeds and all the things that devour our seeds before they can sprout. Where we plant is where we will gather the harvest, if we get one at all.

As a farmer, I have tried to be a caretaker of the soil, a good steward of the land, while helping others to do the same. I have appreciated the recognition of my colleagues, but what matters most, I know, is that I must be a good and faithful servant to the One who cultivates my soul.

Only what we do for Christ will last. To those who will humble themselves and turn from wrong to seek him, God promises that he "will forgive their sin and will heal their land." For thousands of years, humanity has desperately needed that healing. To whom much is given, much is required. We have been blessed with this good earth and its bounty, and we have a huge responsibility to take care of it, to enrich it, to leave it better than we found it.

CHAPTER 6

WHAT GOT YOU HERE WON'T GET YOU THERE

WHAT GOT YOU HERE WON'T GET YOU THERE

The tomato buyer and the squash buyer peppered me with questions as they strolled around Cedar Meadow Farm with a photographer. "How many months of the year do you have living roots in the soil? How many plant families do you have in your rotation? What cover cropping practices do you use?"

The visitors were from the Blue Apron meal kit delivery service, and they had come to see for themselves whether we were doing what I claimed we were doing. Hoping to develop another market for my tomatoes, I had placed a call to the company. This was in 2017, and I had heard that Blue Apron was getting ready to go public. The company seemed to be bursting with ambition, and I saw an opportunity.

Two weeks later, I got a call back—but not from a buyer. It was from a marketer, who didn't know that I had contacted the company. He asked whether I would be willing to educate Blue Apron's farmers on how to grow produce using cover crops and with less tillage. "We're wondering if we might work out some sort of consultation program," he said.

"Sure, let's talk," I said, "and by the way, I'd be interested in selling some of my tomatoes and squash to you, too."

I met with the buyers and marketers in their New York City offices. What motivated me to drive three and a half hours and pay $56 to park a few hours in Manhattan? I wanted to meet these people face-to-face. That's the right way to do business. Get close enough to shake hands with the key decision makers. Doing that has served me well over time.

After the meeting in Manhattan, the Blue Apron folks treated my wife, son and me to a tour of their packing and shipping center in Elizabeth, New Jersey, where scores of workers busily filled boxes with farm produce and recipes that customers had ordered. During our outing, we talked about how I might become both a consultant and a grower for the company.

Blue Apron was particularly interested in our heirloom tomatoes and in a small, sweet butternut squash called honeynut. The company officials seemed aware, whether through consumer surveys or intuition, what their market was demanding. They were eager to demonstrate to their customers that their growers used regenerative agricultural practices.

The fact that the buyers asked such specific questions when they visited our farm spoke volumes about what their customers had been asking them. Blue Apron wanted to document on its website that we and its other suppliers were the real deal. Those visitors were "ground truthing" our claims. No other company has put that much effort into verifying for its customers what we do here. I was impressed by that tenacity, and I agreed that year to supply Blue Apron with tomatoes and squash. This was a company that seemed to be going places.

However, as the company was preparing with much fanfare to begin trading on the stock exchange, Amazon crashed the party: it announced that it was buying Whole Foods for $13.7 billion. The king of online sales was purchasing a national grocery chain popular among millennials who favor organic and sustainable foods.

In other words, an elaborate delivery system was now assured of a plentiful supply of the right stuff, coming at just the right time to stifle any competition. The timing of the announcement, while investors were contemplating Blue Apron's initial public offering, was no coincidence. Whole Foods shares jumped 28 percent on the news—and when Blue Apron debuted on Wall Street in June 2017, its opening price was down by a third from earlier expectations.

That summer, Blue Apron informed me that it was scaling back and had decided that a consulting outreach to farmers would have to wait. And that December, the company informed me that it would not be committing to any contracts with farmers for 2018 but would just buy on the open market. *No way*, I thought. *I can't grow for a company that will not commit to me.*

I figured at the time that Blue Apron was cooked. The company was already facing a parade of competition from others getting into the grocery delivery space. The market had become saturated as they all tried to figure out what people wanted most. Did they want a daily meal service? Did they just want meals for special occasions? Blue Apron was the largest of the lot, but defending that title cost the company a ton of marketing money as it fought to retain customers— an expense that tends to trouble investors.

Online shopping is here to stay. Food is here to stay. It makes sense that online food shopping should be a big deal. In the meantime, Blue Apron and its competitors large and small have been positioning themselves, looking for the right business model. As I was writing this, Blue Apron was hanging in there, going through a succession of CEOs and struggling with losses in its customer base and revenues. Nonetheless, it was still the nation's top meal kit service, and its remaining customers had been putting in larger and more frequent orders, suggesting a foundation of loyalty that could carry it through.[11]

As a farmer, I have been involved with all three of those companies—Blue Apron, Whole Foods, and Amazon. In 2016, I sold two pallets of heirloom tomatoes to Amazon, which was testing the online sales market in two New York City zip codes. Considering that Amazon then proceeded to buy Whole Foods the next year, I'm thinking those tests showed real promise.

I have been selling produce to Whole Foods for fifteen years, ever since I heard that the company wanted to get on the locally grown bandwagon. I met with the regional produce buyer, who seemed unimpressed with my spiel about the way we grew our vegetables. When I mentioned that our pumpkins were cleaner, though,

Beautiful, clean pumpkins that we sell to Whole Foods Market.

---

11   Leo Sun, "Blue Apron Shows Faint Hints of a Recovery," The Motley Fool, Nov. 5, 2019, https://www.fool.com/investing/2019/11/05/blue-apron-shows-hints-recovery-earnings.aspx.

he was all ears. "It's because we use cover crops and no-till," I explained. "The pumpkins grow on the cover crop residue that we roll down before planting."

To this day, we still sell dozens of tractor trailer loads of pumpkins and winter squash each year to fifty-six Whole Foods stores in our region. Recently we have added thousands of packets of a two-ounce tomato variety, about the size of a golf ball, that we dubbed "Aunt Cheri's Tomitas." (It took me a month to convince my wife to use her name for our very own brand.) Whole Foods screens its growers carefully. Each producer recently was asked to fill out a long questionnaire, and soil health was a significant aspect of those questions. Cedar Meadow Farm scored high.

My wife Cheri with Aunt Cheri's Tomitas in Whole Foods Market.

Whole Foods and the others made their mark by doing what smart businesses have always done—staying alert and tapping into trends. In the business world, the survivors are the nimble ones, the resilient ones, who welcome changes as new opportunities. In the internet age, those changes can come day by day. Wise investors understand the oft-repeated warning that past performance is no guarantee of future success. They pay attention to the changing market. Wise farmers should take to heart a similar principle: what got us *here* won't get us *there*.

> # Wise farmers should take to heart a similar principle: what got us *here* won't get us *there*.

A DIZZYING PACE

As we explore how farmers can get to where they need to go, we should look first at where agriculture has been—that is, at what got us here. Our past performance in many ways has been spectacular, but the farming community has also made some big mistakes. Our future success will depend on what we have learned and what we can overcome.

The past century has seen a dizzying pace of advances that have allowed agriculture to feed more people while cultivating less land. Modern agriculture is feeding huge populations and doing so relatively inexpensively. Science has given us so much with new and progressive ways.

At the turn of the last century, America was largely agrarian. About 40 percent of the total U.S. population lived on farms, and most of the rest lived in rural areas. Today, about 1 percent live on farms, and only a fifth of the population is rural. Until about 1940, the nation had six to seven million farms. Then began a steep drop of about a million farms each decade until the decline leveled off to about two million farms from 1980 through today.[12]

The average farm size since 1900 has nearly tripled, from about a hundred and fifty acres to nearly four hundred and fifty acres in 2020. Much of today's production, however, comes from mega-farms with sales of a million dollars or more annually and an average size of nearly three thousand acres. According to recent USDA statistics, farms like that have been producing two-thirds of the total crop value each year on only a quarter of the nation's total farmland. Despite that major share of the production, though, those large operations account for less than 4 percent of the total number of farms in America.

By contrast, small farms with sales of less than $50,000 a year are cultivating about the same portion of the nation's total farm acreage as the mega-farms, but they produce only about 3 percent of the total crop value. Those smaller farms, which tend to be a few hundred acres at most, account for three-quarters of the total number of farms in the country, according to the USDA data.

Meanwhile, crop yields since 1900 have increased dramatically. Wheat farmers today get a yield on average that is more than three times greater per acre. For cotton farmers, it's four times more. For corn farmers, five times more. Such gains have resulted in greater total agricultural production despite fewer farms and less land under cultivation. We also are doing it with a lot fewer workers: USDA figures show that farm labor has been cut by three-quarters since 1948.

However, with that greater productivity, prices have fallen: until the 1950s, corn and wheat prices routinely were three to six

---

12   Sources for section "A Dizzying Pace" include United States Department of Agriculture, "Farms and Land in Farms, 2018 Summary," April 2019, https://www.nass.usda.gov/Publications/Todays_Reports/reports/fnlo0419.pdf; Jayson Lusk, "The Evolution of American Agriculture," Blog, June 27, 2016, http://jaysonlusk.com/blog/2016/6/26/the-evolution-of-american-agriculture; Betsy Freese, "Number of Farms in U.S. Drops as Acreage Size Grows," Successful Farming, Feb. 17, 2017, https://www.agriculture.com/news/business/farms-in-us-drops-size-grows.

times higher than today, when adjusted for inflation. In other words, farmers have more to sell but must sell it for less. The greater supply has benefited grocery shoppers, who get more for their money once those prices, too, are adjusted for inflation.

It all comes down to this: we have a lot less farmland today and a lot fewer farmers, most of whom still have only a few hundred acres. A small minority of the farms have a few thousand acres, and they produce the lion's share of our nation's crops. The average yields are far better than in years past, but real prices that farmers get for their crops today are far lower. And that's the way it is, down on the farm, in 2020.

Much of that change has taken place in the wake of World War II. Think of what was happening in those years. Soldiers were coming home in droves, raising families, and moving into the new suburban subdivisions that once were farm fields. Those were the baby boom years, with a swelling population and all those hungry mouths to feed. The postwar years were also a time of innovation and discovery. New technologies and techniques abounded. America was shooting for the moon.

Agriculture, too, was transformed, with production more than doubling. Along came new synthetic fertilizers, pesticides, and a steady improvement in the quality of seeds. Agricultural machinery became bigger and better. Farming became a science. It was out with the old, in with the new. With a little technical assistance, the amber waves could become a tidal wave.

Or so we thought. Nature has a way of declaring, *Not so fast!* To see that clearly, we need only take another look at the steel moldboard plow, which transformed agriculture during the American march westward, and at the tractor, which in the twentieth century pulled that plow into a new dimension.

PREPARING A BREADBASKET

John Deere's invention of the self-scouring plow in the 1830s soon made the older plows, which couldn't manage the prairie sod, obsolete. The innovation was just what farmers needed to slice up the nation's midsection and, in time, turn it into a breadbasket. It was another example of how technology seemed to have won the day—and how nature would have the final say.

Eight decades later, the John Deere company was at it again. The company turned its attention to the internal combustion engine, which was giving the horse its first real competition. The new technology wasn't coming to the farm all that quickly, though. Various entrepreneurs had been trying their hand at making a marketable tractor, but most of those start-ups failed. In 1911, only about seven thousand tractors were sold nationwide. Nonetheless, John Deere, like any good business, was keeping track of trends. The company began developing a few prototype "tractor plows."

By 1918, the market for the tractor technology was ripening. America was experiencing an agricultural labor shortage as the Great War continued to rage and countless farm boys were fighting "over there" on the front, some never to return. That year, John Deere changed its strategy from development to acquisition, since its own designs had been unsuccessful. The company assessed its competition, picked out a leader, and purchased the Waterloo Gasoline Engine Company for $2.1 million, shifting its resources to the manufacture and sale of the Waterloo Boy tractor.[13]

As demand for tractors grew, farmers from coast to coast began transitioning from horse power to horsepower. Still, tractors didn't

---

13   Neil Dahlstrom, Neil, "The True Story of the Waterloo Boy Tractor," *The John Deere Journal*, March 5, 2018, https://johndeerejournal.com/2018/03/the-true-story-of-the-waterloo-boy-tractor/.

outnumber horses and mules on American farms until 1954.[14]

As is typical with any new technology, some people resist change. They prefer the familiar. In fact, some of the early tractor designs didn't have a steering wheel but instead used "line steering" to give farmers the feeling that they were holding reins.

> Like it or not, the tractor was here to stay. I'm thinking that folks who similarly dismiss today's innovations—online shopping, as just one example—are often getting it wrong, too.

Like all innovations past and present, the tractor had to prove itself to the skeptics. A few of the John Deere corporate officers, for example, were reluctant to embark in this new direction because, after all, the company had risen to prominence by manufacturing implements drawn by horses. Even as the tractor was greatly increasing productivity, some farmers regarded it as a mere fad, soon to pass and more trouble than it was worth. Time would prove them wrong. Like it or not, the tractor was here to stay. I'm thinking that folks who similarly dismiss today's innovations—online shopping, as just one example—are often getting it wrong, too.

The self-scouring plow was a game changer in that it allowed the early prairie farmer to till his homestead efficiently so that he could feed his family. Federal policies, including the Homestead Act of 1862, had encouraged thousands of settlers to pour into the Great

---

14 Bill Cawthon, "From Waterloo to the world," Promotex online, Feb. 1, 2006, http://www.promotex.ca/articles/cawthon/2006/2006-02-01_article.html.

Plains and try their hand at farming. Many of them scarcely knew what they were doing as they tore up the turf, trying to make a living.

Half a century later, the tractor increased the utility of the plow exponentially. Now, great expanses could be tilled with relative ease. The farmers were turning the prairie into a cornucopia for the nation. In the 1910s and 1920s, the rains were plentiful, federal farm policies were generous, and wheat prices were high. The result was another land boom as a new wave of wannabe farmers streamed into the region. By the end of the Roaring Twenties, nearly a third of the Great Plains had been cultivated. On much of what remained of the grasslands, farmers grazed their livestock.

After the stock market crashed in 1929, wheat prices tumbled from $2 a bushel to 40 cents. Hoping for a bumper crop to make up the difference, farmers plowed up even more of the land, believing that the downturn was temporary and that they would soon see a change in the economy. They did. It got worse. The nation was entering a decade of economic and ecological turbulence.

LESSONS FROM THE DUST

For much of the 1930s, a drought parched the Great Plains from Texas into Canada. It was most severe in a region centered on the Oklahoma panhandle of about three hundred by six hundred miles. Deep-rooted prairie grass had once protected the prairie topsoil, but millions of acres of that turf had been plowed under.

The boom years were over. Not only were the crops shriveling in the drought, but prices fell even further. Many of the "suitcase farmers" from back East packed up and went home, abandoning their fields and leaving them barren, at the mercy of the gathering winds. Others fled in desperation in search of work, many of them migrating to California, which turned out not to be the promised land that they had imagined. The "Okie" refugees weren't welcome.

When the wind came sweeping down the plain, as it does so relentlessly in those parts, it carried countless tons of topsoil with it in swirling storms of dust that blocked the sun and settled back to earth in dunes and drifts. Without living roots or plant residue covering the surface, the soil quickly eroded. The Dust Bowl, as it was dubbed by an Associated Press reporter, continued until rainfall returned to normal levels in 1941. It was considered the worst environmental disaster in U.S. history.

By 1934, about 35 million acres already were lost to farming, and the topsoil was rapidly disappearing on 125 million acres more. It was a sizzling year that would remain the hottest on record until 2014. On April 14, which began as a gentle and bright spring day, the winds picked up again and a massive dust cloud—a churning wall of dirt thousands of feet high and hundreds of miles wide—swept across the Plains at high speed, engulfing communities. The widespread devastation that it left on "Black Sunday" focused public attention on the plight of the people and on the need to protect the land that sustains them.

Two weeks after the storm, on April 27, President Franklin D. Roosevelt signed the Soil Conservation Act. By then, the dust clouds were darkening the skies even over the nation's capital. Clearly, this was a matter of national priority. Hugh Bennett, considered to be the father of the soil conservation movement, had long campaigned for measures to protect farmland from erosion. Testifying before Congress in support of the conservation bill, he pointed to a window where the powdery dust that once was Great Plains soil was settling on the sill. "This, gentlemen, is what I've been talking about," he said.

The next year, when the new legislation established the Soil

Conservation Service, Bennett was a natural to become its director. He quickly launched "Operation Dustbowl," focusing attention on the drought's hardest-hit regions. He appointed soil scientist Henry Howard Finnell to lead the effort.

Finnell had already been studying better ways to grow crops in the dry southern Plains. During the 1920s, working in the Oklahoma panhandle, he had concluded in a series of reports that even with normal weather conditions, the typical farming methods of the time would produce good wheat harvests in only four of every ten years, since most of the rainfall never penetrated deeply enough into the soil to benefit the crops.

The trick, he pointed out, was to capture more of the moisture using terraces and contour planting to reduce runoff. He suggested rotating crops and, after each harvest, leaving the stubble as mulch to add organic matter and retain water. Those techniques, he said, would double the likelihood of consistent success.

Nobody paid much attention to those reports when the rains still were coming regularly. The harvest seemed bountiful. Why mess with a good thing? Folks started to listen, though, when the rains stopped, the winds howled, and the topsoil took to the skies.

Even then, though, the conservation efforts were met with skepticism by many farmers wary of newfangled ways. Though the need for a solution was obvious, the prairie farmers didn't necessarily appreciate these newcomers telling them what they were doing wrong and how they should change. "Much of the land could still produce crops," Finnell wrote, "if the farmers would only change their attitudes."

Finnell set up demonstration projects to show farmers how to incorporate the moisture conservation techniques. Other areas, no longer suitable for agriculture, were restored to grassland. The projects were manned by workers from the Civilian Conservation Corps and the Works Progress Administration, established as part of the New Deal to create jobs for the unemployed.

By 1936, about forty thousand farmers were adopting the conservation methods, and they were farming 5.5 million acres. By 1938, even as the drought and the "dusters" continued, the endangered acreage had been reduced by half. The new ways were working. The worst of the hard times was over. A few years later, the rains were returning to the Plains and so were the farmers, who began to get record prices for abundant harvests as the economy geared up for another world war.

A decade later, in the early 1950s, a few years of drought led to more dust storms—but this time the damage was far less extensive. Farmers were continuing to use the conservation techniques introduced during the Dust Bowl years, and the government had permanently restored millions of acres as national grasslands that held the soil firmly in place.

Could a drought as long and severe as the Dust Bowl happen again? In a word, yes—because it has happened before. Studies have established, for example, that prolonged droughts occurred in the sixteenth century and in the tenth to thirteenth centuries and that they likely were longer and drier than the drought of the 1930s. The region, which is also known as Tornado Alley, is prone to swings of climate and wild weather.

Through the millennia, the prairie grass endured wind and drought, but it was helpless against the plowshare. Agriculture didn't cause the rains to stop, nor the winds to blow, but it did cause the dust to rise. The lesson from those years was that improved crop management and a progressive approach to tilling could control erosion, preserve moisture, and improve yields.

TURNING TO NO-TILL

In the decades since the Dust Bowl, farmers increasingly have recognized that the tilling itself has been a major part of the problem. Though it helps to plow judiciously, it's better to plow sparingly, if at all.

Across the country and around the world, farmers have been steadily switching to no-till and low-till methods of planting that leave the turf intact, and they have been incorporating cover crops to enrich the soil and further protect it from runoff and erosion. Though many still resist changing their ways, these methods are hardly newfangled. They are more like a return to the old ways, from a modern perspective.

The no-till concept caught public attention with the publication of *The Plowman's Folly* by Edward H. Faulkner in 1943. "The fact is that no one has ever advanced a scientific reason for plowing," the author declared, echoing what soil scientists had been saying about the risk of depleting the soil of its natural nutrients. In his own test plots, he had transformed the soil and produced higher yields by planting without plowing and incorporating "green manure crops"—that is, cover crops. This was a man ahead of his time. I was struck, when reading his book recently, by the parallels between what he was saying eighty years ago and what I am advocating today.

In the early no-till experiments, researchers found that the need for weed control was a major drawback. Without a plow to till them and kill them, the weeds flourished. It wasn't until after World War II, with the development of powerful herbicides, that anyone took the no-till idea very seriously, but farmers still were in no hurry to give up the plow.

ARTIFICIAL ENRICHMENT

As synthetic fertilizer became widely available after the war, farmers began to develop a chemical dependency. It had been around for a while—by the early 1900s, scientists understood how to produce ammonia and nitric acid as its basic components—but it was expensive. Even if they could have afforded it, farmers lacked machines to apply it and had little guidance on using it effectively. For the most part, farmers continued to fertilize their fields with manure, or with legumes that they plowed into the soil.

With the Second World War came a demand for nitrogen to use in explosives. New factories began pumping out massive quantities of ammonia to meet that requirement. When the war ended, the demand for explosives was replaced with the need for food supplies in the United States and abroad. At the same time, chemists learned to produce phosphates on a large scale, and massive deposits of potash were discovered around the world. The fertilizer industry, supported by extensive research funding, was on its way.[15]

That is how agriculture got hooked on chemicals. The synthetic fertilizers, along with the development of a wide variety of pesticides, made farmers' labors shorter, their yields higher, and their profits greater. They could farm more acres and make a better living off the land. For that same reason, farmers had welcomed John Deere's improvements to the ancient moldboard plow. That's why they had retired their horses and climbed onto tractors. And that is why they began crisscrossing their fields with sprayers behind those tractors.

Before the artificial enrichment of the chemical revolution,

---

15   Gary Hergert, Rex Nielsen, and Jim Margheim, "Fertilizer History P3: WWII Nitrogen Production Issues in Age of Modern Fertilizers," Cropwatch, University of Nebraska–Lincoln, Institute of Agriculture and Natural Resources, April 10, 2015, https://cropwatch.unl.edu/fertilizer-history-p3.

farmers routinely planted a variety of crops and rotated them. Today, many farms grow just one crop or two, the most profitable ones, year after year, acre upon acre. The soil is treated as if it were simply a medium to hold the roots in place for the growing season. The reasoning seems to be that whatever nutrients the plant needs can simply be applied. The crops may thrive, but the soil dies.

THE POWER OF DIVERSITY

Man does not live by corn and soybeans alone. Our bodies, like the soil itself, need an array of nutrients, derived from diverse sources, and we ignore that principle to our peril. A healthy ecosystem requires crop diversity, but agricultural policy has supported a few cash crops at the expense of others.

> Before the artificial enrichment of the chemical revolution, farmers routinely planted a variety of crops and rotated them. Today, many farms grow just one crop or two, the most profitable ones, year after year, acre upon acre.

When crop insurance programs cover only corn and soybeans, for example, farmers understandably will think twice before taking the risk of planting a small grain such as oats. They will continue planting those two crops for years and decades, and the result is barely better than a monocultural crop system.

Introducing a small grain to the rotation helps. Adding several is better. The concept of growing three cover crops together was one I discovered from Dr. Abdul-Baki, the USDA researcher whom I introduced in Chapter 1. I was proud to take that revolutionary idea to the upper Midwest during a speaking engagement in Bismarck, North Dakota, on a bone-chilling January day of 17 below zero. That's when I met Gabe Brown, one of the world's most influential farmers in inspiring regenerative practices. Also on the program was Jay Fuhrer, a USDA soil-health specialist who reported on his research using a whopping eight species of cover crops.

In a recent year at Cedar Meadow Farm, we were growing twenty-seven species, some of them cover crops and some cash crops. (One year I planted forty-three cover crop species at once, but that was more for bragging rights.) Such diversity undeniably improves soil health and resilience. Crops are less vulnerable in both the dry years and the wet ones. They are less likely to fail.

No-tilled corn planted into a diverse mix of cover crops at Cedar Meadow Farm.

Diversity is nature's idea of crop insurance. God put us here in a garden of flourishing plant life, nourished by the soil. Those plants in turn protect and enrich the soil, and they take care of one another's differing nutrient needs. It's a beautiful balance of give-and-take. Wise gardeners respect and preserve that balance. They understand that when the soil is permitted to take care of itself, it will take care of them, and their families, and anyone who partakes of what grows in it. As the Bible says: "Let the fields be jubilant, and everything in them."

> **Diversity is nature's idea of crop insurance.**

We have not always been good gardeners, even when we have meant well. Some of the government programs that seem to do farmers a favor could end up turning them into wards of the state. Though subsidies have played an essential role in hard times, farmers have demonstrated that they have the ingenuity and resources to do quite well without a handout.

Consider the experience of New Zealand and Australia. In the 1980s, both governments abruptly eliminated many of their crop insurance and subsidy programs. Some farmers went out of business, but those who survived reinvented themselves. No longer beholden to the government, they relied instead on their natural innovation. Instead of focusing on just wheat, wheat, wheat, they began growing several crops, in rotation.

I've been to Australia twice. Most of the farmers there today have as many as a dozen grain bins instead of the two or three that once was typical there, and as we see here in the U.S. That's because they grow a diverse species of cash crops to build a healthier soil. Farmers now can feel reasonably assured of a decent harvest whether the rainfall is sufficient or scarce.

The Australian farmers also are depending on cover crops to build organic matter, which retains more moisture. One Australian farmer, Josh Walters, visited Cedar Meadow Farm between my trips to his country. He was amazed at our rainfall, which was double the 20 inches per year he averages at his farm. He invited me to speak at a field day that he hosted at his farm. "We need to build a bigger bucket," he told me while I was there. "We need to capture every drop of water that falls from the sky and keep it in our soil. That is our crop insurance."

A monocultural system simply does not produce the healthiest food. If it is a given that diversity is desirable, then it also is a given that government policies that discourage diversity have not served us well. Sure, a one-crop farm is possible when you put the soil on life support—but will such a business model be sustainable as the market changes and consumers demand that the agricultural industry respect and protect nature and be more in line with its ways?

Though yields are higher than ever overall, the environmental concerns are troubling. One example among many: in the Gulf of Mexico, off the coast of Louisiana where the mighty Mississippi flows to the sea, is a "dead zone" the size of New Jersey. It results from nitrogen and phosphorous that wash into the river, eventually feeding massive algae blooms that deplete the oxygen in the Gulf waters. Much of that chemical pollution comes from agricultural production in the heartland of America. Our planet has numerous such dead zones. This one is the largest.[16]

Make no mistake: there will be consequences for agriculture's chemical dependency and loss of diversity. The people are speaking

---

16   "Gulf of Mexico 'dead zone' is the largest ever measured," National Oceanic and Atmospheric Administration, Aug. 2, 2017, https://www.noaa.gov/media-release/gulf-of-mexico-dead-zone-is-largest-ever-measured.

out loudly that they want something better, and their dollars are doing the talking. They sincerely care. Farmers who ignore the consumer trends will find their customers switching to competitors who demonstrate a commitment to diversity and environmental responsibility.

BUILDING ON THE PAST

It's not that farmers are greedy or oblivious to the damage that agriculture can cause to our soil, our waterways, our wildlife, and our health. Like all of us, farmers wish to make a living, and in doing so, they also are feeding massive populations worldwide. If they are to consider giving up the plow and the fertilizers and the pesticides, they need an alternative. They need something to replace the tools that got them to where they are today. They can't just go cold turkey.

Here's the good news: even without those crutches, farmers still can get to where they need to go. They do have alternatives. Those who embrace them and use them effectively will thrive as they progress into the twenty-first century. In nature, the survivors are the adapters. As their environment shifts, they adjust to it. Eventually they find themselves transformed. It doesn't happen overnight. But it happens.

Our strength will be our willingness to make amends for our mistakes and find new and better directions, all the while remembering where we have been. Those who hold to the old ways often view progress with suspicion, sometimes contempt. Those who insist on the new ways sometimes are contemptuous of tradition. The first approach is regressive, the second is reckless. The best approach is to keep the best of the old ways and apply that foundation of wisdom to bold new ideas. Humanity is endlessly creative, but we grow by building on the past, not abandoning it.

Once, bald eagles were nearly extinct. Today I see them regularly at our farm. Agriculture has made progress toward turning around the damage. We have learned some important lessons on how to get along with nature—and we have many more, I'm sure, yet to discover. Much of our progress will come from paying close attention to what nature is telling us, as farmers have done since antiquity. The learning never ceases.

No-till planting winter squash into a rolled cover crop of hairy vetch.

It's not easy to turn away from what has seemed to work for so long, but farmers can start by taking baby steps before advancing to a full stride. By trying their hand at cover crops, along with no-till planting and rotation, they can see for themselves how effective the alternatives can be. They will find success when they consider cover crops to be as essential as their cash crops, with a plan in place from planting to harvest and for all the steps along the way.

A big opportunity awaits farmers who work in concert with

nature, not against it. As we have seen time and time again, we cannot tame nature, and we dare not ignore it. What we must do is understand it, respect it, and learn from it as we go about our business of farming sustainably and responsibly. Our mission must be to regenerate the land while farming profitably. That's the goal— and if we do this right, we'll get there.

CHAPTER 7

BETTER TO EDUCATE THAN REGULATE

BETTER TO EDUCATE THAN REGULATE

"It takes a lot of oysters and crabs to send a kid to college these days," James "Ooker" Eskridge, the mayor of Tangier, told me as we toured his island in the middle of the Chesapeake Bay—and I knew just what he meant. After all, it had taken a lot of tomatoes and squash and other produce for me to do the same. Any farmer understands the labors that go into earning a living, whether it's off the land or out of the waters.

I was visiting the island in 2017 to meet up with Ooker and other watermen and ask a simple question: *Are things getting better?* For decades, nutrient pollution from a variety of sources, but much of it from farmland, had been compromising the Bay, the largest estuary in the United States and third largest in the world. For a few years, studies had been indicating some improvement—but what did the people who depended on the Bay to earn a living have to say? Many had been forced out of business.

As I talked to the mayor, a film crew followed us for a documentary called *Living Soil*. Produced by the Soil Health Institute, the documentary examined in part the relationship between agriculture and the decline of the Chesapeake and the cooperative efforts under way to save it. (You can watch the documentary at www.livingsoilfilm.com.)

Touring the island in a golf cart, we began talking about how the farmer and the fisherman have a lot in common. For one thing, they are both to a great extent at the mercy of the weather and the elements. Government regulations restrict what they can do and when. The farmer needs productive soil, the fisherman needs productive water. Both play an indispensable role: "The farmers, like the watermen, we're feeding the world," Ooker said.

Farmer meets fisherman: me with Ooker, the mayor of Tangier Island in the middle of the Chesapeake Bay.

The Chesapeake indeed was healthier than in years past, the mayor told me. I felt gratified that my advocacy of cover crops and no-till farming had been contributing to that improvement. Ooker now felt confident about sending his daughter to college. The oysters and crabs were returning, and they were paying her way.

I understood that feeling. In our family, the Tillage Radish had paid the way. That innovation of mine, widely planted as a cover crop throughout the Chesapeake watershed, not only generated a lot of tuition money but also played no small role in beginning to restore the Bay.

"Where is your daughter going to college?" I asked. He told me that she was a student at Liberty University in Virginia.

"Really? My daughter, Lauren, just graduated from there," I said. We had even more in common than I had thought. Here was a man who, like me, depended on a healthy environment to make a good living. Here was a father who, like me, wanted the best for the next generation—and whose daughter had chosen the same school as my daughter. We were two men comparing notes about life and our dreams for our families, and we were finding that we weren't all that much different.

I had met Ooker in 2013, when I invited him to one of my annual field days at Cedar Meadow Farm. I felt that farmers would benefit from getting to know a Chesapeake fisherman whose way of life had been changing partly because of agricultural practices upstream. It often helps to put a face to an issue. When he spoke at the field day, Ooker told the farmers that the people of Tangier appreciated our efforts to do something about the alarming condition of the Bay. His message was one of cooperation, not condemnation, but he also made it clear that the pollution was threatening the waterman culture that for generations had depended on an abundant harvest from the Chesapeake.

Only several hundred people live on Tangier Island, which is part of Virginia, nineteen miles off the coast. The island is about a mile wide and three miles long, and its highest point is only about four feet above sea level. A century or two ago, the island was at least twice that size, according to archaeologists. Though scientists have been warning that global warming and rising sea levels will eventually swallow the community, most of the islanders don't buy that explanation—and the fact remains that the land has been dwindling away for hundreds of years. The prevalence of stone arrowheads that still often wash up during storms suggests that the island once was a good hunting ground and far larger than it was even at the time of the first European settlement.

The erosion of the shoreline remains a major issue, with significant acreage slipping into the Chesapeake each year. On the western side of the island, the erosion stopped after a seawall was built there in the late 1980s. The islanders have been working for years to procure funding to build jetties and seawalls around the rest of the island to slow the loss of the land mass, but such proposals tend to get lost in the bureaucracy.

"It's just a common-sense thing," Ooker told me, pointing out that farmers and fishermen both understand the need to preserve the land that provides their livelihood. "You would want to protect the topsoil because it's vital to farming, like our shoreline is vital to us."

Some fear that without intervention, the islanders one day will be forced to abandon their community—and with them will go their distinct and venerable culture. Tangier Island is among the last of the waterman communities in the Chesapeake. The crabbing and oyster industry, along with summer tourism, sustains virtually all the households on the island. In the early 1800s, the island was a center for Methodist revival camps, and the community remains deeply religious. The island's relative isolation—it takes at least an hour by boat to get there—has helped to preserve a unique way of life and a dialect found nowhere else on earth. Some say it evolved from the eighteenth-century brogue of the early Cornish settlers, with Tidewater influences.[17]

It was four years after Ooker visited my farm that I returned to Tangier Island for the filming of the documentary. The consensus

---

17 Stephen Blakely, "Tangier Island, Va.," Soundings, June 16, 2017. https://www.soundingsonline.com/features/tangier-island-va.

was that the Bay clearly was improving, although it had a long way to go. During the long boat trip out to the island, I chatted with the captain. "I hear a lot about pollution in the Chesapeake," I said, without telling him why I was visiting. "You're out here every day, right up close to the water, and you see what's going on. Have you noticed any changes?"

"Oh, for sure," he said. "The water's cleaner. The fishermen are happier."

Ooker, too, confirmed that the Bay had been improving and the islanders were feeling more hopeful. They had noticed a healthier growth of seagrasses. The blue crab population was growing, and the watermen were bringing in considerably more oysters than the paltry few bushels that a day's labor had gotten them several years earlier. This was not the abundance that earlier generations had experienced, but it was marked improvement.

> **The consensus was that the Bay clearly was improving, although it had a long way to go.**

Two years later, in September 2019, I revisited Tangier Island and the mayor. I was among about a dozen farmers from the Pennsylvania No-Till Alliance that the Chesapeake Bay Foundation invited to the island to assess how the situation there had changed. We tried our hand at dredging for oysters, and we set out crab traps, returning for the catch the next day. I found it fascinating to experience the ways of the watermen.

We also tested the water for turbidity and nitrates, and the results supported what the locals had observed: slowly but surely, the Bay is turning around. As in our previous encounters, the fishermen expressed their gratitude for what the farmers in the watershed had been doing to cooperate in the cleanup. After years of education on farming methods that would help to protect the Bay, we were making an impact. We had a long way to go, but we had turned a corner.

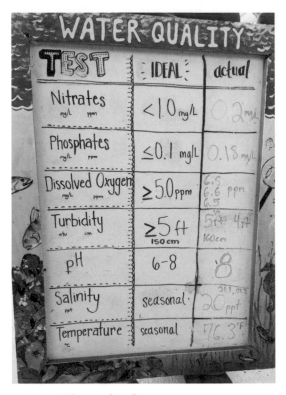

The results of our water testing
in the Chesapeake Bay.

Though Tangier Island feels far removed from the world that I know in southern Lancaster County, we are separated by only about 140 miles as the crow flies. We are neighbors, really, and our destinies are intertwined.

Lancaster County is the bullseye for the Chesapeake Bay's nutrient pollution issues. Yes, the practice of agriculture in my county has been only part of the problem, but it's a big part. Other sources include runoff from development, industry, sewage plants, and farmland elsewhere in the watershed, but make no mistake: much of the nitrogen and phosphorus and sediments that have damaged the Bay originated in the wide expanses of farmland in beautiful Lancaster County.

Most of us who farm these fields are long past the denial stage. The finger-pointing, for the most part, is a thing of the past. The evidence is clear, and we recognize the truth: the chemical fertilizers and manure from decades of agricultural operations have found their way downriver to the estuary, flushed from the soil with every rainfall, and the damage is done.

The question now is what we will do about it. A few years of improved conditions must not lull us into thinking that our mission is accomplished. In truth, we have only just begun. As farmers we can expand our use of cover crops and retire more of our plows as we turn to no-till. Those two steps alone will have much to do with whether the fishing culture of a tiny island can survive.

GETTING THE MESSAGE

I have been hearing about the Chesapeake Bay issues for virtually all my farming days. Even back in the early 1980s, when I was a young farmer just out of high school, I heard the talk about preventive steps that we should be taking or else we would be required to take them.

That didn't come across as a threat but as a statement of fact. The message was clear: if farmers didn't learn to clean up their act, they could face government restrictions such as how much fertilizer they could apply and how much land they could plow. There wasn't much talk at first about cover crops. The emphasis was more on no-till.

The major concern, initially, was more the phosphorus runoff than the nitrogen. A primary source of that runoff is the manure from livestock operations, whether from cattle in Lancaster County or the chicken industry on the eastern shore of Maryland.

For two years in the mid-1990s, Cedar Meadow Farm participated in a phosphorus monitoring program. We were chosen for the program because I had been a strong advocate of the advantages of no-till, and the study sought to compare the runoff from farms that plowed to those, like mine, that did not. A local agriculture company, Brubaker Agronomic Consulting Service,[18] was selected to conduct tests by setting up a series of devices that trapped the water running off our fields, and then measured the amount of sediment in those samples. By its nature, phosphorus adheres tightly to soil particles, so it tends not to leave the farmland unless the soil goes with it.

Though the consultants did find some phosphorus dissolved in the water, they found next to no sediment in the traps at our farm. The reason: our fields had not been tilled. Undisturbed by a plow, the soil stayed firmly in place, keeping much of the phosphorus with it. The consultants told me that Cedar Meadow had the best results of all the farms in their study. The no-till farms, they concluded, were doing a better job than the others of keeping phosphorus out of the waterways.

When the government began to require conservation plans, such as how many gallons or pounds of manure they could spread per acre, many farmers resisted, but eventually most came to understand that the regulations were not too onerous. Farmers, by and large, recognize that the regulations have helped the environment—and that they can even save money through their greater awareness

---

18   Mike Brubaker, the founder of Brubaker Agronomic Consulting, went on to become a Pennsylvania state senator from 2007 to 2015. He worked with farmers, government officials, and environmental groups in implementing legal requirements for the Chesapeake Bay watershed. Senator Brubaker represented the farmers well, and his influence was appreciated by the agriculture community.

of how to manage the manure so that it benefits the soil instead of washing into the waterways.

When farmers recognize that something worthwhile is happening out in their fields, something tangible, they want to keep up the good work. Most live by the creed of "do no harm"—and because cover crops and no-till have served them and their neighbors well through the years, I find it easier today than a generation ago to promote the benefits.

The evidence that Lancaster farmers are getting the message is clear in the landscape: less than a third of the county cropland is still tilled, and virtually all the farmers who have switched to no-till also plant cover crops. They understand that it's the right thing to do. They know that those methods have made them better farmers. Not everyone is on board, of course. Some worry about the cost of conversion, and some simply remain unaware. Lancaster County still has a lot of furrowed fields.

SO FAR TO GO

I know firsthand, from my travels, that the agricultural community has a long way to go. The statistics do seem impressive. Across the United States, according to the Soil Health Institute, farmers in 2018 were using cover crops and other soil-health practices on 18 million acres. Other projections have found that cover crop acreage nation-wide was increasing by about 15 percent every year and was expected to reach about 20 million acres by 2020.[19] That's twice as much as a decade earlier. But it's still less than a tenth of America's cropland.

Despite the progress, the regulators still have the farmers in their

---

19 "Cover Crops and Carbon Sequestration: An overview of cover crop impacts on U.S. cropland carbon sequestration," Sustainable Agriculture Research & Education (SARE), July 2018.

spotlight. If we don't do better, we can expect further restrictions and requirements. Particularly in this era when we hear so much about global warming and climate change, people worldwide generally will be seeing their governments taking a stance. In other words, we will all likely be facing more rules on what we can and cannot do.

I'm not interested in wading into a political quagmire. My aim, always, is to bring people together, and to harmonize their thinking, not polarize it. For some, the big question is whether climate change is happening; for others, it's why it is happening. Scientists, politicians, economists, theologians, businesspeople, and just about everybody else have been weighing in on the issue. We need all those perspectives, because it's fair to say that the truth is somewhere in the mix.

As for me, I'm interested in common sense. I'm interested in what is happening out in my fields, and in the fields of my fellow farmers around the world—and what happens in the waters downstream from us, and in the atmosphere above us. I can say this: something's up. It's hard to deny that many regions of the world have been experiencing a shift, whether it's due to normal and primordial weather cycles or the result of what mankind has visited upon the land, air, and sea.

As a man of the land, I am striving to do my part for what God has placed in our care. He gives us knowledge and discernment, so we must not summarily dismiss what a lot of smart folks are trying to tell us. If we accept, for example, that greenhouse gases endanger the earth's atmosphere, then it stands to reason that we should be doing something about it.

To whatever degree humanity has influenced climate change, a primary contributor could be leaving the soil uncovered. When I was a barefoot boy, back in the tillage days, I couldn't walk in a cultivated field of tomatoes on a sunny summer day. The soil was too hot. I have measured the temperature of bare soil at 130 degrees Fahrenheit while a few feet away, where a cover crop grew, the soil temperature was 50 degrees cooler. That's the dramatic difference that covering the soil can make—yet at any given time, half the world's 3.3 billion acres of cropland is uncovered, according to statistics cited by Dr. Christine Jones, a noted soil ecologist from Australia with whom I've crossed paths over the years.

Consider what such high temperatures can do to the microbes in the soil. Like humans, microbes are at their best with temperatures in the 70s. If we keep them happy, they will work for us and provide biological benefits, including natural fertilizers. Mycorrhizal fungi, for example, help to translocate minerals such as phosphorus directly into a plant's roots.

Hot soil wastes water as well. If the ambient temperature is 95 degrees and the soil surface is 130 degrees, the moisture will be sucked out of the soil into the air. Water vapor is a greenhouse gas, and it has increased significantly over the last century. NASA's website includes a study titled "Water Vapor Confirmed as Major Player in Climate Change."[20] Unprotected soil contributes to that cycle. Heat and moisture radiating back into the atmosphere will certainly influence the climate one way or another.

Carbon dioxide is often blamed for a changing climate, even though it represents only 0.04 percent of the atmosphere—and only a fraction of it comes from fossil fuels. Greenhouse growers often boost yields by pumping in carbon dioxide, which is essential for plant growth and a healthy ecosystem. I'm not advocating pollution, of course. We must be responsible for what we can control—and the release of carbon dioxide into the atmosphere is something the

---

20 "Water Vapor Confirmed as Major Player in Climate Change, " NASA, November 17, 2008, https://www.nasa.gov/topics/earth/features/vapor_warming.html.

farmers can help to control. When the soil is disturbed by tilling, carbon mixes with oxygen to produce the gas.

I learned that from Dr. Don Reicosky, a now-retired USDA soil scientist. He conducted sophisticated research in which he built a special chamber to capture carbon dioxide emissions from tilled soil. In his presentations, he has used a slide that depicts a farmer plowing a field with flames rising from the ground. During the 1990s and beyond, his research influenced many farmers to park the plow and keep carbon in the soil where it belongs. Dr. Reicosky has been among the many who have participated in my annual field days at Cedar Meadow Farm, and he gave me his slides to use in my own presentations.

As I ponder the political messaging focusing on excess carbon dioxide as the culprit for the world's wacky weather, I must pose this question: could it be that the fossil fuel emitters get so much blame because they are easier to vilify? We must consider other sources of the gas. Covering up a few billion acres of bare cropland could go far toward stabilizing the weather. In a no-till and cover crop world, we could have vast expanses covered with diverse plant life growing year-round in undisturbed soil.

That is how our planet was designed to function. That is how farmers might take the lead in doing something about greenhouse gases. Cover crops have immense potential to keep carbon dioxide out of the atmosphere by sequestering it in the soil. And no-till agriculture, by leaving the soil undisturbed, also keeps the carbon where it belongs. Farmers like practical solutions, and those are a couple of good ones waiting for us to make better use of them.

One study concluded that cover crops planted on 20 million

acres would offset the emissions from 12.8 million cars.[21] I manage only about three hundred acres, but it's all no-till with a very diverse crop rotation—and I let my cover crops grow longer than is typical. If my calculations are correct, my farming methods on that amount of land have been offsetting carbon dioxide emissions each year by as much as what three hundred cars would produce. And I like to think that my influence globally has been far greater than that as I have promoted carbon sequestration through the years. Recently I have been trying to verify how close to carbon-neutral my farm is—and I'm aiming for carbon-positive.

"The soil is naked, hungry, thirsty, and running a fever," says Ray Archuleta, a top advocate for soil health. He has led many farmers to that epiphany moment that changes their mindset about how they work the land. Ray has spoken at several of my field days and has an unforgettably engaging way of connecting with an audience. And he walks the talk. My family visited his home when he lived in North Carolina, where he had a

> "The soil is naked, hungry, thirsty, and running a fever."

large no-till garden thriving with cover crops. Today he has a farm in Missouri and has incorporated sheep into his cropping system.

Like Ray, I have spent years advocating a better way through no-till and cover crops. We must pay attention to what nature is telling us. Where agriculture has caused harm to the environment, we must own up, and those progressive practices go far toward making amends.

---

21 "Cover Crops and Carbon Sequestration," Sustainable Agriculture Research and Education (SARE), accessed May 11, 2020, https://www.sare.org/Learning-Center/Topic-Rooms/Cover-Crops/Ecosystem-Services-from-Cover-Crops/Cover-Crops-and-Carbon-Sequestration.

This goes beyond making amends, however. Agriculture can also be a tool to bring great benefits to the environment. We are already seeing how cover crop farmers, as they enrich the soil, are helping to regenerate the Chesapeake and other waterways. We know now that they are also contributing to a healthier atmosphere. And no matter where you stand on what causes global warming, that cannot be anything but good. There's really no downside.

The trouble with regulations is that they tend to draw battle lines. In day-to-day practice, some of them don't make much sense. They can be downright annoying, and yet they are not without purpose. The intent is to pursue lofty goals that should be in the public's best interest. They do something else, too: the prospect of more rules incentivizes farmers to take the necessary steps to keep the regulators at bay. In doing so, they have seen the benefits to their land and to their wallets and naturally want more of what works.

The bottom line is that the regulations, or the threat of them, have played into Lancaster County's remarkably high adoption rate for cover crops and no-till. So has the influence of the dozens of advocacy groups for the Chesapeake. I know, though, that the biggest factor has been the fact that our farmers are smart citizens with a sincere desire to do what's right. We do indeed have a long way to go, but we have come so far already—and let's give credit where credit's due.

I took this picture from an airplane the first week of December to prove that 65 percent of Lancaster County, Pennsylvania, uses cover crops.

I'm not one to do anything to get awards. These come as a result of my passion and desire to be an environmentally responsible farmer.

BETTER A CARROT THAN A STICK

Because many farmers today consider cover crops to be the only way to go, some don't even take the subsidies that are available as an incentive to grow them. For something that you know is essential to making a good living, you need no incentive. You're going to do it anyway.

Nonetheless, various funding sources dangle the carrot of subsidies in hope of coaxing more farmers to get moving. It's certainly a gentler approach than the stick of regulations and mandates. Subsidies help to address a major barrier to cover crop adoption: the added expense required to get started. Farmers like the concept but balk at the up-front costs. They wonder whether it makes sense, from a business perspective, to wait for that return on investment. I understand. I wondered that same thing when I ventured into cover crops back in the mid-1990s. The question persists in the farming community.

To address that concern, farmers can receive subsidies in various forms to reimburse them for certain costs involved in planting cover crops. The state of Maryland, for example, levies a tax on people's sewer bills—folks like to call it the "flush tax"—and uses the proceeds to pay farmers to plant cover crops. The federal Natural Resources Conservation Service (formerly the Soil Conservation Service) has some provisions for funding cover crops as well. Other opportunities are available in some parts of the country, often depending on the extent of concern over the local watershed. The funding is generally a combination of government and nongovernment sources.

In my region, Chesapeake Bay advocacy groups have often paid for costs of cover crop seeds and planting, and money also has been available for manure management. After a quarter century of publicity, farmers tend to be well-aware of those opportunities. At the local level, some municipal officials have recently investigated the prospect of paying farmers to grow cover crops as a means to offset soil loss or degradation from subdivision and other development plans.

Subsidy programs typically seek to get farmers started with funding for the first three years or so, although some programs, such as Maryland's, don't have that limit. The general goal is to get farmers accustomed to the concept. That can be an effective stimulus. After those first few years, though, the initial funding should become unnecessary. As farmers integrate cover crops into their budgets, they see what they are gaining and give up the crutch.

How many farmers in Lancaster County take any kind of subsidy to plant cover crops? I have asked that question to numerous people who work with the NRCS, with the Lancaster Conservation District, and with various Chesapeake Bay organizations. Most agree that it's fewer than 15 percent. Nationwide, surveys indicate that more than 60 percent of cover crop farmers bear the entire cost.[22]

In Europe, cover crops are mandated more or less depending on which country. If farmers there wish to partake in their government's subsidies, which they virtually all do, then they must maintain a cover crop from September to mid-November in any field that doesn't have a cash crop growing in it. In the United States, no such mandate exists, to my knowledge. So far, the American way is to use the carrot instead of the stick. Regulations have their place, but we would rather encourage than require.

---

22 "Cover Crops Survey Analysis: Cover Crop Report Documents Yield Boost, Soil Benefits and Ag Retailer Roles," Sustainable Agriculture Research and Education (SARE), accessed May 11, 2020, https://www.sare.org/Learning-Center/Topic-Rooms/Cover-Crops/Cover-Crop-Surveys/2013-14-Cover-Crops-Survey-Analysis.

THE SMART THING TO DO

When I am out on speaking tours across the country, folks often ask me what it's like living and farming near the Chesapeake Bay, which is perceived far and wide as a model for what should be done in other watersheds that are facing their own environmental challenges. "We've heard about all those regulations," they tell me. "How do you deal with that?"

I tell them that I'm not motivated by the regulations. I'm motivated by a desire to improve the land. Whatever I do is for the sake of the farm. The regulators have certainly recognized my conservation efforts, but I do not manage my farm to please them. I manage it the way I do because I care about my land—and that is what pleases them. They know I care about the environment. I demonstrate it day by day.

Farmers are the ultimate environmentalists. How could they be otherwise? The environment is their workplace, after all. It's their place of business. They don't generally define themselves as environmentalists, though, because the word has somehow become associated with bureaucrats and meddlers who would limit their freedom—and farmers value freedom. They are wary of regulations and limitations, not because they don't want to do their part for the environment, not because they don't care, but because they care deeply. They care about preserving the spirit of independence.

Farmers also value common sense, and they have a pretty good sense of smell when it comes to detecting anything that threatens their liberty to think for themselves. They don't want orders. They want explanations. Once they understand why, they can be trusted to do what's right. Once they see the benefits, they are more than willing to act.

That's where I come in, as a consultant, speaker, mentor, and author. I explain the why and the how, and my fellow farmers take it from there. As good businessmen, they want to do what makes sense, whether the government requires it or not. Farmers balk at regulations, but they are eager for education. As they learn the value of cover crops and begin planting more of them, two things will happen: the environment will continue to benefit immensely, and fears of regulation will subside.

Market forces, not the government, will effectively mandate the planting of cover crops. Farmers will require it of themselves as they go after big opportunities in new markets. They will see for themselves why it's the smart thing to do, both environmentally and financially. With that attitude, they can go far. They will proceed proactively, not begrudgingly. They won't be laboring under the dispiriting message that they'd better clean up their act or face the consequences. Instead they will seek to get in on the action to reap the benefits.

There's no better motivator for a farmer to adopt a new practice than a "value added" component. I have heard many times how "brilliant" my gourds are—a result of them not touching the ground but being grown on a rolled cover crop in no-till conditions.

Education is the answer. Farmers are primed to learn about cover crops, and they are talking to one another about their efforts. Doing this right requires skills and experience, and many are actively pursuing what it takes to succeed by asking others what has worked for them. They are listening, discovering, and spreading the word.

Together we are learning that we all profit from farming methods that regenerate our land, water and air and that regenerate the livelihood of those who depend on a thriving environment. We can do well by doing good. The watermen of Tangier Island would agree that we must do what we can to protect what sustains us. As the mayor pointed out, it's just a matter of common sense.

> We can do well by doing good. The watermen of Tangier Island would agree that we must do what we can to protect what sustains us.

CHAPTER 8

CONNECTING WITH THE CONNECTORS

CONNECTING WITH THE CONNECTORS

Though I was eight days too late to meet Tío Gabriel and thank him for inviting me to Argentina, I felt the man's spirit in his widow's tearful hug. Never will I forget that scene: his grieving loved ones greeted me, a stranger far from home, as if I were family. I knew that here, too, I was home.

This was late in November 2019—that's springtime south of the equator. I had flown over five thousand miles to participate in a weeklong *Suelos y Sistemas* (Soils and Systems) conference in Argentina and Uruguay, organized to help growers, agronomists, and others deepen their knowledge of regenerative agriculture. My Argentine hosts, Sandro Raspo and his wife, Maria Teresa, greeted me at the airport in Buenos Aires and, during the course of my stay, drove nearly two thousand miles to escort me to various venues where I helped to spread the word on cover crops and no-till.

"How did you hear about me?" I asked my hosts as we walked through the airport. I was curious as to how their organization had learned about this Pennsylvania farmer and his passion for cover crops.

"Oh, it was all Tío Gabriel," Maria Teresa said in fluent English. "That's Sandro's uncle. He organized everything about your visit here, every stop, every field day. He told us he'd seen you on a dozen YouTube videos. 'We must get him to come to Argentina,'

159

he kept saying. He was certain you would have much to teach us about cover crops."

"Great!" I said. "I'm really looking forward to meeting him." The man certainly sounded like a kindred spirit.

"Well, he was eager to meet you, too, Steve—but that's not going to happen. Tío Gabriel had a heart attack last week. He's gone."

I struggled for words and found the only ones that seem to fit such moments: "I'm so sorry."

"Yes, and he was only sixty-three," Maria Teresa said, pausing to reflect on the loss. "But you will be meeting his wife and their son and daughter. They knew how much Tío Gabriel wanted to meet you, so they're coming out the day after tomorrow to one of the farm field days that he had organized."

Everywhere we traveled, I met extraordinarily friendly people. I felt like a rock star. My hosts had my name embroidered on their shirts, and folks were asking me to autograph their hats and the brochures for the event. Film crews followed us, and I was interviewed several times for various agriculture news outlets. I appreciated the opportunity to communicate the thing or two that I've learned about good farming in my days.

On my third day there, we pulled up to a farm to join fifty or so people already gathering to hear me give a talk and to tour the cover crop plots together. "Come over here," Sandro said as I stepped from the car. "We want you to meet my uncle's family. This is Olga." Softly sobbing, his widow opened her arms and hugged me very tightly. No words were needed, but I did recall a phrase from my limited Spanish. "Sólo Dios sabe," I whispered to her. "Only God knows."

As our somber group stood in that field, there were tears all around in memory of Tío Gabriel and all he had meant to so many. It was a profound moment of deep connection for this farm boy from so far away. Though we did not all know each other's language, we spoke the universal language of a shared vision to become better caretakers of the land. As it gives to us, so must we also give back to it.

Me with Sandro, Olga (Gabriel's widow, center), and her two children in Argentina.

STRONGER TOGETHER

On the pampas of Argentina and Uruguay, vast expanses of soybeans grow where once the fabled gauchos herded cattle. In the mid-1990s, the grasslands were quickly converted to cash crops, largely due to the advent of Roundup Ready soybeans.

Though cover crops are new to many farmers in the regions that I visited, the farmers are keenly interested in how the concept can help them deal with a rising water table—despite twenty-five years of weather data showing that precipitation has been relatively the same. No longer does the pampas grass pull water from the soil year-round. In its place are the cash crops, mostly soybeans but also corn and wheat, which grow only three or four months of the year.

The farmers are learning that cover crops can accomplish what the prairie sod can no longer do. With new roots in the soil through the year, the water table can stabilize. Cattle can also graze on the cover crops—and many farmers are evaluating the importance of reintroducing cattle into the rotation to further promote healthy soil.

As I talked with the growers there, my thoughts turned to another farmer far away, my friend Monte Bottens, who grows corn and soybeans in Illinois. He took an early interest in converting to cover crops, and then began grazing cattle on his land, a highly unusual move out in corn and soybean country. Monte launched a business called Grateful Graze to direct market the meat. "We believe that soil health, animal health, and human health are linked together in ways that we are just beginning to understand," he explains on the company website.

"Monte, why did you do such a crazy thing as bringing cattle onto your corn and soybean farm?" I asked him during an interview for one of my online webinars. I did about a hundred and fifty of them for CoverCropInnovators.com, a website that I established and

In my office doing a cover crop webinar.

recently sold to Lessiter Media so I could dedicate more time to new ventures. The webinars are on the general topic of cover crops and soil health, with a variety of guests joining me for the discussions.

"Why? Because that's what the customer wants," he answered, explaining that the market was increasingly interested in meat from livestock that can graze outside in its natural habitat. Consumers are willing to pay a premium price for that quality. By paying attention to market trends and acting on them, Monte has future-proofed his operations—and he is enjoying the new challenge. He has made farming fun again.

I met Monte in 1999 when he was a crop advisor in California, where he lives part of the year to run one of his agriculture businesses. I was scheduled to give a talk at a conference on the virtues of cover crops and no-till. As I drove two hundred miles from Sacramento down to the San Joaquin Valley, I realized that not many farmers there had bought in to the concept. I didn't see one cover crop or no-till field.

Returning in 2006, I met up again with Monte, and we visited a farm that in the intervening years had begun planting cover crops. We went out to the fields with the farm owner, Alan Sano; the farm manager, Jesse Sanchez; and Jeff Mitchell, the University of California specialist who had originally invited me there. I suggested that we shovel a sample of the soil to check its texture and smell. Digging in for a good look and sniff has become kind of my trademark.

"Oh, here's an earthworm," I mentioned casually as I saw one wriggling in a soil sample. The farmers were ecstatic. You'd have thought they had hit the lottery: they were jumping for joy, literally, and gathering around this worm to take pictures of it. Alan told me it had been thirty years since he'd seen earthworms in his fields. They were amazed at the difference that cover crops had made.

*Alan Sano of Firebaugh, California, with
"the first earthworm I found in thirty years!"*

Through the years, in my travels to teach about cover crops and no-till, I have met with Monte at his Illinois farm and talked with his neighbors in his very modern farm shop. He in turn has become a trusted consultant for other farmers, and in his own operations he has taken a step beyond what I have done by adding cattle grazing to his management. I know he would agree with me that it is an honor to serve as a mentor. We have something essential to share with those who care to listen.

I am gratified whenever anyone who has worked with me moves on to another level. A generation ago, I felt like a lone voice. Cedar Meadow Farm was an "island of sustainability," a documentary by PBS declared in 1999. The documentary, *Land of Plenty, Land of Want*, was a segment of the *Journey to Planet Earth* series narrated by

Matt Damon. Back then, I was one of a very few who were teaching about cover crops. Today I am one of many voices in a movement that has branched out in intriguing new directions. To paraphrase Scripture, we are stronger together than individually, in the manner that a braided rope is stronger than a single strand.

I see my role as a connector who pulls people together for a common purpose. I am proud to have helped give birth to this movement, and I will be prouder still when nobody needs to hear what I have to say anymore because they already are putting it into action. I know of nowhere in the world, though, where we couldn't do better.

Meanwhile, I will continue to do my part in spreading the word, whether I am speaking at a conference to an audience of a thousand, a field day on the other side of the world, or to a few local Amish farmers in a barn. I am confident that those who hear me will tell others what they have learned. When farmers mentor other farmers, good things grow from it.

BRANCHING OUT

At Cedar Meadow Farm, we have been devoting acreage to the production of CBD hemp since the 2018 Farm Bill and the Pennsylvania legislature approved the crop. In Chapter 2, I described our new venture. It's only natural, then, that my speaking engagements also have branched out to focus on how to grow hemp with cover crops and no-till.

Across the nation, I've met many resourceful farmers who are exploring opportunities in this rapidly growing enterprise. They are recognizing not only the growing market for CBD products but also the public demand for responsible agriculture. These entrepreneurs want to know how the methods that I have long advocated can be applied to hemp.

Together we are fine-tuning the systems that work best, the varieties that grow best, and the conditions necessary for an optimal yield. Much of the learning is in the doing, and the learning never ends. At Cedar Meadow, we had a good growing season for our first hemp planting, but we were intent on finding out how to do even better the next year. Hemp is a strong plant that isn't very susceptible to disease and insects, and neither caused us much concern. We discovered that hemp does well when planted into a cover crop such as Dutch white clover, which grows low and slow. Hairy vetch works particularly well as a cover crop preceding the planting of hemp. They have a natural synergy.

Above all, we demonstrated conclusively that hemp farmers do not need to till the ground and lay down plastic, as has been the typical practice. They are far better off with no-till and cover crops— and so is the soil, as is the case with anything else that farmers choose to grow. Same idea, different crop. In fact, hemp is proving to be ideal for pairing with cover crops.

Finding my attention increasingly focused on the use of cover crops for growing hemp, I launched a new website, www.hempin-novators.com. Hemp Innovators is a community of farmers and professionals navigating the hemp industry together. I established this educational forum after interested farmers repeatedly told me that they could find no reliable resources for large-scale commercial production. I realized that farmers need the right training if they are to take advantage of this major opportunity. They need to join and build a network.

"I MET THIS FARMER ... "

Whenever I ask people who have succeeded with cover crops how they got their education, most say something like: "Well, I met this farmer ... " They might not use the term "mentor," but that's what they mean. They got good information when they needed it most from a source they trusted.

Farmers interested in cover crops should count on nothing less. Those who have blazed the trail and proved their skill are eminently qualified to guide others along the path. Experienced cover croppers are natural mentors who can offer solid advice on following best practices and avoiding common mistakes. In my public appearances, I have often emphasized the importance of those personal relationships. Farmers tend to prefer that one-on-one approach.

French farmer Frédéric Thomas is one such colleague with whom I've had a longtime rapport. I met him in 2008 during my first excursion to Europe, when I was invited to speak in his country and in Germany. I think of Frédéric as my French brother because we have so much in common—same age, same size farm, and for years even the same model John Deere 6400 tractor. We both are educators. He founded the organization Biodiversity, Agriculture, Soil & Environment (BASE), a network of farmer groups in Europe with well over a thousand members and a mission that parallels my efforts. He is also editor of *TCS* magazine, which he founded in 1998. His website, www.agriculture-de-conservation.com, is a wealth of ideas and information (and easily translated from French with Google Translate).

Moreover, Frédéric and I think alike. I consider him a mentor, and we keep in touch just about every month. Each in his own way, and with his own audience, we are helping other farmers to learn and thrive as good stewards of the land. That's the power of connection.

When farmers think of one another as neighbors, whether next door or across the ocean, great things happen. Farmers have always come up with plenty of good ideas as they stand talking at the fence. Only difference today is that the fence is defined differently.

> Farmers have always come up with plenty of good ideas as they stand talking at the fence. Only difference today is that the fence is defined differently.

Besides mentors, a wealth of information is available from other sources. "YouTube is my mentor," one grower proclaimed after I'd finished a presentation. And he had a good point. Online videos and webinars offer valuable guidance on cover crops. Agronomists and other educators are also good sources of information, and many of them have practical farm experience as well.

"How do you stay on top of everything you do, Steve?" folks sometimes ask me. I tell them it's the team around me that helps me focus on what I do best. When I attend conferences, I learn a lot from the agenda—but often the most valuable thing that I take away comes from the people I've met there. I'll chat awhile, ask a question or two, and invariably someone tells me that I need to talk to so-and-so. And when I do, so-and-so tells me that I also should talk to someone else—and so on. I'm always on the lookout for those who can teach me more than I think I know. They can become valuable team members.

Mentors need their own mentors, who will have their own mentors. That's how knowledge expands and circulates. In Argentina, for example, the farmers informed me that they get higher protein levels in soybeans by planting the crop after a legume instead of after cereal rye, as I have long recommended. They had invited me there to show them some new tricks, but they also showed me one that I certainly must investigate. We became, in effect, mentors to each other. One must always keep an open mind.

Success requires the initiative to reach out and make such connections. You can do that with a laptop, certainly, but I'd rather start with the handshake. Either way, farmers will advance the cover crop movement by networking, which is just another way to say they should get to know folks, ask how they've been doing, and look for anyone trying to do it better. In other words, they must connect with the connectors. Nothing unusual about that. Farmers have operated that way for ages.

Farmers listen to farmers. After all, they are in the foxhole together, facing the same assortment of perils. They are familiar with the extremes of the weather and of the market. Farmers far and wide face differing issues, of course—too little water, too much, a scorcher, a freeze, pests and disease—but many of the challenges of working the land are universal. That's why I can connect not just with farmers in Lancaster County but also with farmers coast to coast, around the world, and on both sides of the equator. All of us are in the business of making the most of the soil. It's human nature to trust those who know what you're going through.

And so, farmer to farmer, trust me when I say this: we should be listening, too, to those who don't necessarily know what we go through. We must pay attention to our customers, even those who have never set foot on a farm. They are telling us, with their purchasing power, something critical that we need to hear. They care about how we grow our crops.

That's not a threat to our livelihood, unless we choose to ignore them. It's an exciting opportunity. What our customers buy is what makes us thrive. They are telling us loud and clear what they want—and as increasing numbers of my fellow farmers are recognizing, it is in our own best interest to give it to them. Ultimately, that's the most important connection of all.

CONCLUSION

THE SEEDS
FOR SUCCESS

THE SEEDS FOR SUCCESS

Strolling the streets of Thessaloniki, Greece, I admired the view of mighty Mount Olympus, fifty miles in the distance, which the ancient Greeks believed was the dwelling place of their gods. I knew, too, that the apostles Paul and Timothy must have beheld that same snow-capped peak two millennia ago as they spread the gospel of Jesus throughout these gently rolling hills along the shores of the Aegean Sea. In my hotel room, I opened my Bible to read Paul's epistles with new eyes. Here I was steeped in history, religion, and tradition.

I was also steeped in agriculture. The purpose of my four-day stay in early February 2020 was to consult with farmers at the biennial Agrotica trade fair, an international show of two thousand exhibitors of agricultural machinery, equipment, and materials. And there, too, the modern and the traditional mingled. Among the 150,000 attendees, I saw men in the black monastic robes of the Greek Orthodox Church examining new models of John Deere tractors and collecting agricultural brochures detailing the latest technological advances.

My mission at the exhibition was to identify farmers, particularly cotton growers, interested in no-till and cover crops. I had been invited to the show by Roian Atwood, the sustainability director for Wrangler, whom I had met at a conference and had interviewed for one of my webinars. He was looking to expand the company's "Rooted Collection" of sustainably produced jeans, which was proving successful in America, into the European market.

Wrangler's Rooted Collection blue jeans. Note the cotton to make these jeans was from Alabama.

Inside pocket of Wrangler's Rooted Collection jeans. Note the farm the cotton was grown on.

Roian's initiative is a prime example of the growing corporate recognition that good farming is good business. At the Agrotica fair, he wanted to learn whether individual farmers recognized it, too. He wanted to find cover crop farmers who wanted to participate in this new market. My role would be to gauge their interest, evaluate equipment availability, and strategize ways for farmers to grow cotton in a sustainable fashion. I would meet with those farmers, answer their questions, and encourage them.

This was a bustling event attracting people from forty-nine countries, with most of the attendees from southern Europe. I was pleased to get such an opportunity to spread the word on regenerative agriculture. Before I talked with any of the farmers, though, Roian instructed me to make the rounds of the show to identify no-till planters and other equipment that I might be able to recommend. Instead I saw dozens of plows and disks and other tillage equipment. Clearly the farmers in this region would need more than good advice on cover crops and no-till. They would also need the right tools, readily available, and they would require training on how to use them properly.

That concern was top of mind among the prospective cover crop farmers with whom I met privately, with the help of an interpreter. They wanted information about how to get started and the proper methods to use. A big question, however, was this: *How will we be able to do the planting? There are very few no-till planters in this part of the world.*

Supply rises to meet demand. I've seen that happen elsewhere through the years as I have urged farmers to turn from tillage to better ways. When the desire is there, the tools appear. I'm confident that the equipment manufacturers will rise to the occasion. If they determine that their customers in southern Europe are sufficiently interested in no-till and cover crops, the right kind of machinery is sure to show up. After all, it's readily available in many parts of the world now. Meanwhile, my job, wherever I travel, is to encourage that interest. My mission is to shift the movement into high gear wherever possible.

After I returned home from the trade fair, a no-till farmer from Greece contacted me after seeing one of my social media posts about the trip. He told me that he, too, had been a no-till farmer for a couple of years. He had not been at the show, but he wanted me to know that a group of like-minded farmers had been working for a generation to raise awareness in Greece. And a few days later, a university professor there contacted me to say that he and his colleagues were interested in being a part of the cover crop initiative to produce cotton for Wrangler. No doubt about it: the seeds had been planted. Now it was just a matter of growing them.

Around the globe, on every continent where I have consulted with farmers, I have encountered that same mindset for success. I have heard their questions in a wide variety of languages and accents and dialects, including the familiar one of my Lancaster County home. Some of the men and women I meet are new to the world of no-till planting and cover crops, but they are generally eager to know more. They understand what's at stake, and they want to do right by the environment.

I often ponder the question, How can these wild berries bear fruit without any fertilizer or pesticides applied? Daughter Lauren and son David in the background "foraging" for wild berries.

Sowing seeds of success includes educating the leaders of tomorrow. I'm showing our granddaughter an earthworm—her very first soil health lesson!

And they want to do right by their families, too. Once, the prevailing question among farmers was: "Can we afford to do this?" Now, the question is becoming: "Can we afford *not* to do this?" Consumers have made it clear that they will pay well for farm products that have been kind to the soil. They prefer nature's ways. Major corporations recognize that demand, and farmers, increasingly, are seeing it too. Our future depends on how we respond. Today's decisions determine tomorrow's success. In the years ahead, family farms could flourish or perish.

One fine autumn day, as the leaves were just beginning to fall, I went for a walk in the woods. The soil was soft and rich, yielding a profusion of vegetation. *No nutrient deficiencies here*, I thought. *Looks like nature has things well under control.*

Nobody dumps chemical fertilizers in the woods to make all that growth possible. Nobody plows up the landscape to plant the wildflowers and grasses and ferns. This diversity of life thrives instead on the nutrients bequeathed by previous seasons. It's the cycle of growth and decay, generation after generation, that nourishes and protects the life-giving soil.

That's the lesson we must take from the forests to the fields. Our customers increasingly expect farmers to follow nature's example, and we must expect it of ourselves. In doing so, we will be preserving not only this good earth but also the way of life that farmers cherish. We are on our way to a bountiful harvest, but we must do the hard work to get there—never forgetting the divine truth that we will reap what we sow. To future-proof our farms, we must plant now the seeds for success.

ACKNOWLEDGMENTS

Throughout these pages I mention many people who have educated and spurred me on to pursue innovations. I want to acknowledge a few others who have been fellow travelers on this journey.

As you may imagine, I'm not a fan of some of our current U.S. agriculture policies—but I've tried to do something about it. Twice I've been to USDA headquarters in Washington, DC, at the invitation of Rob Myers, north-central regional director of sustainable agriculture extension programs, who coordinated efforts for some key agriculture stakeholders to meet with top-level decision makers. We have recently had such good soil-health advocates as USDA undersecretary Bill Northey, whom I'd met a couple of times at cover crop meetings in Iowa. Jimmy Emmons, a farmer who hails from Oklahoma, has been appointed as south-central regional director of USDA's Farm Production and Conservation division. Our paths have crossed often, and we share a passion for soil stewardship and thriving farms. Having such folks in positions of influence gives me hope that our country will remove roadblocks to long-term regenerative agriculture.

And then there is Rick Bieber from north-central South Dakota, who farms over 10,000 acres in a 16-inch precipitation zone. Despite those who say that cover crops won't work in dry areas or on large farms, he has been quite successful and coined the phrase "Soil Care" to describe the core tenet of his soil-health system. I had heard Rick's message for years, and we finally connected on the phone as I was driving home one day. Interest in growing food packed with nutrients is generally focused on vegetables, but as we talked, I realized that here was a large-scale farmer doing the same thing as me except with broadacre crops. I pulled off to the side of the road, and we talked for a half hour about how to convince other farmers that this was the future. I finally met Rick at the Menoken Farm in North Dakota the following year when I was speaking at a summer field day.

Me with Jay Fuhrer and Rick Bieber evaluating
a field's soil health in North Dakota.

In 2013, a group of five leading farmers from South Africa visited my farm. I kept up with Egon Zunckel, with whom I felt a special connection. I wasn't sure why until I saw a Facebook video of Egon in his tractor cab—and in the background I could hear Bethel worship music, the kind I play on my bass guitar on our church's worship team. In September 2018, at the invitation of Egon and the No-Till Club committee, I was keynote speaker for a three-day conference in Bergville, South Africa. My wife, Cheri, went with me. During a safari that Egon and his wife, Gaynor, arranged before the conference, the four of us forged a friendship. The following March, Egon and I participated in Zsombor Diriczi's series of field days in Hungary. Our wives joined us on that trip, too, cementing our friendship as couples—you might say the trip was good for our soul health as well as for soil health. Afterward we toured Budapest together. Cheri and I look forward to the day when Egon and Gaynor can visit our neck of the woods.

Finally, I have people who ask me, "How do you stay on top of everything you do, Steve?" I usually reply that "it's the team I have around me that helps me focus on what I do best." Megan and Matt Steinruck, a husband-and-wife team who founded Big Picture Studio, have played a key role in the past several years of my outreach. Matter of fact, it was Matt who suggested the title of this book. They help with social media and website design and play a key role in the overall direction of what I do. It's nice to work with good people. It's even nicer when those people become friends.

ABOUT THE AUTHOR

Steve Groff is a farmer, researcher, agricultural consultant, speaker, and author who specializes in the use of cover crops, no-till planting, and regenerative agriculture methods. His speaking and consulting engagements have taken him across the United States and Canada and to such international destinations as Bulgaria, Hungary, Australia, Romania, Belgium, France, Germany, South Africa, Argentina, Uruguay, and Greece. He designed the first roller crimper for cover crops in North America, developed the cover crop Tillage Radish, started the first company in the world exclusively selling cover crop seed, and was the first commercial vegetable farmer to transplant tomatoes with no-till. He has consulted for Wrangler, Stroud Water Research Center, the Chesapeake Bay Foundation, the Natural Resources Conservation Service, Rodale Institute, Penn State University, the University of Maryland, and Cornell University.